E. Greene Charles

Trusses and Arches

Part III

E. Greene Charles

Trusses and Arches
Part III

ISBN/EAN: 9783337349462

Printed in Europe, USA, Canada, Australia, Japan

Cover: Foto ©berggeist007 / pixelio.de

More available books at **www.hansebooks.com**

Graphics for Engineers, Architects, and Builders:

A MANUAL FOR DESIGNERS, AND A TEXT-BOOK FOR SCIENTIFIC SCHOOLS.

TRUSSES AND ARCHES

ANALYZED AND DISCUSSED BY GRAPHICAL METHODS.

BY

CHARLES E. GREENE, A.M.,

PROFESSOR OF CIVIL ENGINEERING, UNIVERSITY OF MICHIGAN.

IN THREE PARTS.

I.

ROOF-TRUSSES: DIAGRAMS FOR STEADY LOAD, SNOW, AND WIND.

II.

BRIDGE-TRUSSES: SINGLE, CONTINUOUS, AND DRAW SPANS; SINGLE AND MULTIPLE SYSTEMS; STRAIGHT AND INCLINED CHORDS.

III.

ARCHES, IN WOOD, IRON, AND STONE, FOR ROOFS, BRIDGES, AND WALL-OPENINGS; ARCHED RIBS AND BRACED ARCHES; STRESSES FROM WIND, AND CHANGE OF TEMPERATURE; STIFFENED SUSPENSION BRIDGES.

PART III.—ARCHES.

EIGHT FOLDING PLATES.

SECOND EDITION.

SECOND THOUSAND.

NEW YORK:

JOHN WILEY AND SONS,

53 EAST TENTH STREET,

1894.

PREFACE TO PART III.

THE curved lines of arches are pleasing to the eye, and may often be introduced with advantage in constructions. Ar. may furnish, under some circumstances, a very economical way of spanning an opening; and arched ribs are employed in other cases, at conspicuous locations, where beauty of design is regarded, or where ample and uninterrupted space beneath a roof is desired. Stone arches have been built for many centuries; at the present time, wood, iron, and steel are also used as materials. If the principles which enable one to ascertain the forces acting in all parts of an arched structure are clearly understood, designs of this type will be more common than they now are; and it is with the desire to do what he can toward shedding some light upon this subject, as well as to give the ability to intelligently design an arch to those who are not familiar with the higher mathematics, that the author submits the following pages to the public.

Most persons experience difficulty in mastering the principles which govern the action of an arch, as they have hitherto been presented. Even one who has successfully worked through the mathematical theory, as he finds it in the text-books, may sometimes lose sight of the actual meaning of each step in the process; so that there is a certain mystery about the application of the formulæ to a specific example, although one may feel confident that the results are reliable. To many constructors a treatise on the arch, as usually written, is a sealed book, and the whole subject is veiled in obscurity. Empirical rules, copying of existing examples, and guesswork have been the refuge of many. While such practice may answer for masonry structures, where the factor of safety as regards *strength* is very large, the introduction of iron skeleton structures, where the pieces occupy definite lines of force, and the sharp rivalry for economical disposition of the material, render a better practice desirable. It is hoped that the graphical method developed in the following pages will enable the reader to understand as clearly the effect of applied forces on an arch, as it has, through the explanations of Parts I. and II., enabled him to analyze trussed roofs and bridges.

From the *bending moment*, *direct thrust*, and *shear*, here obtained at successive sections of the arched rib, the stresses in the chords or flanges, and bracing or web, are derived as if the structure were a simple truss. In finding the resultant stresses in the pieces, the method of Part I. will sometimes be preferred to that of Part II. So far as possible, the formulæ of the text have been obtained by direct and easy ways; and, while it has been convenient to arrive at some of the definite results by the use of the calculus, such results have been obtained from the diagrams, and can in all cases be verified by the reader, for any specific example, by the most simple means.

After the subject is once mastered, the resulting formulæ and applications will, naturally, alone be referred to in working out designs: the author has therefore thought it best to place the results, &c., in direct connection with the explanatory statements, and to have the analytical or mathematical demonstrations follow in smaller type. One who simply desires working-material may omit the matter printed in small type, without losing any of the facts, but must then take some statements for granted.

A distinctive notation for the figures, introduced in Parts I. and II., — capitals for structures and moment diagrams, small letters for the shear diagrams, and numerals for the stress diagrams, — has been generally adhered to. While an acquaintance with Parts I. and II. will aid the reader in understanding more readily the graphical constructions here given, it has been the aim of the author to enter sufficiently into detail to make this part intelligible by itself: hence a few explanations repeated here.

It is believed that many things offered in these pages will be new to most readers. The work is almost entirely the result of independent investigation. A portion of the material was once printed in the "Engineering News," but it has been entirely revised since that time: over one-half of this part is now in type for the first time. The device of increasing the breadth of the parabolic rib, or the thickness of the flanges, from the crown to the springing, while the depth remains constant, — which device will be found in Rankine's "Civil Engineering," — enables the summation of ordinates to be made across the span, as for a beam, rendering the treatment simple. On the other hand, the depth and breadth of the circular rib are supposed to be constant, and the summation is made along the curve. Herein the treatment differs from that of some authors. It is shown that the direct thrust on a right section is not equal to the product of the horizontal thrust by the secant of the inclination of the rib at the section to the horizon, as some writers assume, unless the equilibrium curve is parallel to the axis of the rib. Other points of difference in treatment and result will be found by readers who are familiar with the literature on this subject. The discussion, in Chapter VIII., of the action of the wind on an arched roof, will, it is hoped, be found timely and serviceable; the effect of change of temperature, and the change of form under stress (Chapter XI.), are often ignored by writers; an example of a stone arch of considerable magnitude is worked out in detail; the methods of stiffening suspension bridges are discussed and compared: on some of these points very little has heretofore been given.

C. E. G.

ANN ARBOR, MICH., July, 1879.

TABLE OF CONTENTS.

CHAPTER I.

GENERAL PRINCIPLES.

SECTION.		PAGE.
1.	Arches	15
2.	Funicular Polygon applied to a Curved Rib	15
3.	Relation between Equilibrium Polygon and Bending Moments	16
4.	Value of Bending Moment	17
5.	Remarks	18
6.	Condition to determine H; Invariability of Span	19
7.	Formula for this Condition	20
8.	The Equilibrium Polygon determinate	21
9.	Deflection of the Rib	21
10.	Another Value for Bending Moment	21
11.	Combined Effect of Bending Moment and Direct Force	22
12.	Reversal of Figure; Movement of Rib from Equilibrium Polygon	23
13.	Equilibrium Polygon for a Single Load	23
14.	Direct Force and Shear at a Right Section	24
15.	Sign of Shear; Maximum Bending Moment at Point of Zero Shear	25
16.	Treatment of Arch with Fixed Ends requires Three Conditions	26
17.	First Condition	27
18.	Second Condition: Change of Inclination between Abutments equals Zero	27
19.	Third Condition: Deflection between Abutments equals Zero	28
20.	This Condition not applicable to Hinged Rib	28
21.	Remarks: Abutment Reactions; Shear, &c.	29

CHAPTER II.

ARCH HINGED AT THREE POINTS.

SECTION. PAGE.
22. Three-hinged Arch 32
23. Formula for H 33
24. Stone Arches 33
25. Example 34
26. Caution 35
27. Relation between Equilibrium Polygon and Curve 36
28. The Parabola the Equilibrium Curve for a Load Uniform horizontally 36
29. Suspension-Bridge 37
30. Equilibrium Curve for Partial Load 38
31. Suggested Examples 38
32. Extent of Load to produce Maximum Bending Moment . . . 39
33. Braced Arch 39
34. Shear; Temperature 40

CHAPTER III.

INTRODUCTORY TO PARABOLIC ARCHES.

35. Parabolic Arch 41
36. Vertical Deflection of an Inclined Beam 41
37. Application to Arches 42

PARABOLIC RIB HINGED AT ENDS.

38. Equilibrium Polygon for Single Load 43
39. Proof of Formula 44
40. Formula for Horizontal Thrust 45
41. Computation of y_0 and H 46
42. Remarks 47
43. Computation of Bending Moments 47
44. Table of Bending Moments 48
45. Interpolation 49
46. Examples 50
47. Numerical Value of M 52
48. Shear Diagram 52
49. Shear on a Normal Section 54
50. Formula for Vertical Shear 54
51. Computation of Shear 55
52. Remarks on Shear 55

SECTION. PAGE.
53. Table of Shears 56
54. Extent of Load to produce Maximum Bending Moments and Shears, 56
55. Resultant Maximum Stresses 57
56. Example of Flange Stresses 59

CHAPTER IV.

PARABOLIC RIB WITH FIXED ENDS.

57. Values of Ordinates 60
58. Value of First Equation 61
59. Values of Second and Third Equations 62
60. Solution of Equations 63
61. Remarks 63
62. Computation of Ordinates y_1 and y_2 . . . . .
63. Formulæ for H, P_1, and P_2 . . . .
64. Computation of Values . . . .
65. Computation of Bending Moments . . . .
66. Table of Bending Moments . . . . [illegible]
67. Example 67
68. Table of Shear 69
69. Extent of Load to produce Maximum M and F 69
70. Comparison of Ribs; Fixed and Hinged at Abutments 71

CHAPTER V.

CHANGE OF TEMPERATURE.

71. Action of Change of Temperature 72
72. Change of Span influenced by Material and Cross-section of Arch . 73
73. Formula for H from Change of Temperature 74
74. Application to Parabolic Rib Hinged at Ends 74
75. Formula for Change of Span deduced analytically 75
76. Application to Fixed Parabolic Rib 76
77. Comparison of Arches under Change of Temperature 77
78. Shear from Change of Temperature 78
79. Diagram for Vertical Shear 78

CHAPTER VI.

CIRCULAR RIB WITH HINGED ENDS.

80. Circular Rib to be of Uniform Section 80
81. Experimental Verification 81

SECTION. PAGE.
82. Semicircular Arch with Hinged Ends; Value of y_0 81
83. Segmental Arch; Value of y_0 82
84. Proof 83
85. Formula for H; Value of Ordinates 84
86. Numerical Computation of M 84
87. Shear at any Right Section 85
88. Shear Diagram 86
89. Distribution of Load to produce Equilibrium 87
90. Effect of Change of Temperature 87
91. Shear from Change of Temperature 88

CHAPTER VII.

CIRCULAR RIB WITH FIXED ENDS.

2. Values of Equations of Condition 89
93. Special Values for Semicircular Rib 90
94. First Equation of Condition 91
95. Second Equation of Condition 92
96. Third Equatior of Condition 93
97. Reduction of Equations 93
98. Values of H, &c. 93
99. Table of y_0, y_1, and y_2, for Semicircle 94
100. Example 95
101. Practical Application 96
102. Limiting Position of Equilibrium Curve 97
103. Model as Hinged at Three Points 99
104. Model as Hinged at Abutments 100
105. Effect of Change of Temperature 101
106. Maximum Stress determined by Length of Ordinate; Rib of Rectangular Section 102
107. Rib of Two Flanges 104
108. Rib of Circular Section; General Construction 104

CHAPTER VIII.

ARCHED RIBS UNDER WIND PRESSURE: HORIZONTAL FORCES.

109. Wind Pressure on an Inclined Surface 106
110. Form of the Equilibrium Polygon; Vertical Component of Reaction 107
111. Rib Hinged at Three Points 108

SECTION. PAGE.

112. Value of Bending Moments 108
113. Parabolic Rib Hinged at Abutments; Formula for x_0 . . . 109
114. Proof of Formula 110
115. Another Proof 111
116. Formulæ for H_1 and P 112
117. Shear and Direct Stress 112
118. Shear Diagram 113
119. Circular Rib Hinged at Ends 114
120. Formulæ for H_1 and P 115
121. Experimental Verification 115
122. Parabolic Rib Fixed at Ends; Formulæ for x_0, x_1, and x_2 . . 116
123. First Equation of Condition 116
124. Second and Third Equations of Condition 117
125. Formulæ for H_1 and P 119
126. Circular Arch Fixed at Ends . . . .
127. First Equation of Condition . . . .
128. Second and Third Equations of Condition . . . .
129. Reduction . . . .
130. Formulæ for H_1, &c.; Semicircular Arch . . . . . . [illegible]
131. Sign of Bending Moment 123
132. Example of Normal Forces 124
133. Finding the Reactions 125
134. Equilibrium Polygon; Bending Moments 127
135. Equilibrium Polygons for Vertical and Horizontal Components . 128
136. Shear and Direct Stress 129
137. Vertical Shear Diagram 129

CHAPTER IX.

STONE ARCHES.

138. Location of Equilibrium Curve determines Thickness of Voussoirs, 131
139. Intensity of Pressure 132
140. Circular Arch; Load for Equilibrium 132
141. Limiting Angle for Arch-Ring without Backing 134
142. Example; Data 135
143. Calculations for Steady Load 135
144. Equilibrium Curve for Steady Load 137
145. Calculations for Rolling Load 138
146. Increase of Bending Moment at Crown; Required Depth of Keystone 139
147. Increase of Bending Moment at Haunch 140

SECTION. PAGE.
148. Influence of an Additional Load 141
149. Increase of Bending Moment at Springing; Maximum H . . 141
150. Final Dimensions of Arch 142
151. General Remarks 143
152. Exaggeration of Vertical Scale 144
153. Elliptic Arch 144
154. Example 145
155. Treatment for Horizontal Forces 146
156. Catenary 146
157. Transformed Catenary; Example 147
158. Construction 149
159. Many-centred Arch 149

CHAPTER X.

STIFFENED SUSPENSION-BRIDGES.

160. Necessity for Stiffening 150
161. Inverted Arch 150
162. Horizontal Girder 151
163. Distribution of Rolling Load between Cable and Truss . . 152
164. Comparison of Inverted Arch and Horizontal Girder . . . 153
165. Horizontal Stiffening Girder Hinged at Ends only . . . 154
166. Stiffening Girder of Varying Depth 154
167. Ead's Arch, or Lenticular Stiffening Girder 155
168. Bowstring Stiffening Girder 157
169. Fidler's Stiffened Suspension-Bridge 158
170. Ordish's Suspension-Bridge 159
171. Erect and Inverted Arch combined 159

CHAPTER XI.

BENDING MOMENTS FROM CHANGE OF FORM.

172. Displacement from Bending Moments 161
173. Displacement and Bending Moments from Compression . . 162
174. Parabolic Rib Hinged at Ends 162
175. Remarks; Example 164
176. Displacement from Change of Temperature 165
177. Initial Camber for Arch 166
178. Parabolic Rib with Fixed Ends 167

SECTION. PAGE.

179. Circular Rib Hinged at Ends 168
180. Analytical Discussion 170
181. Circular Rib Fixed at Ends 172

CHAPTER XII.

BRACED ARCH WITH HORIZONTAL MEMBER; OTHER SPECIAL FORMS; CONCLUSION.

182. The Usual Analysis not applicable 173
183. Change of Span from Stress in a Piece 174
184. Stress in a Piece from H and P 174
185. Formula for H 176
186. Application of Method 178
187. Example; Stresses from H and P . . . .
188. Computation of Tables . . . .
189. Values of H . . . .
190. Diagrams and Table of Stresses for Equal Cross-Sections . .
191. Sections proportioned to Stresses 182
192. Bracing with Vertical Struts 186
193. Cast-Iron Arch as a Breast-Summer 186
194. Gothic Rib for Roofs 187
195. Remarks on Designing 189

1

ARCHES.

CHAPTER I.

GENERAL PRINCIPLES.

1. **Arches.** — An arch may be considered to be any structure which, under the action of vertical forces, exerts horizontal or inclined forces against its supports or abutments. Such a definition will include not only the roof of two simple rafters, but also the suspension bridge; and we see no objection to so including them. The case of two rafters we need not touch upon: the suspension bridge only comes incidentally within the scope of this part, until we take up the means of stiffening such a structure under a moving and partial load.

2. **Funicular Polygon applied to a Curved Rib.** — Suppose that a curved rib A C E B, Fig. 1, of any material which possesses stiffness, for instance iron, is attached by a pin, on which it can turn freely, to each of the points of support A and B, and has suspended from it certain known weights, represented by W_1, W_2, &c., at known points. The weight of the curved rib itself is not at present considered. The rib, if flexible, as a cord or chain is flexible, will tend to assume the shape of the funicular, or equilibrium polygon, proper to these weights in their respective positions. If we lay off the load line 2–1, to any scale, space off on it the weights in succession, assume any convenient point 0, draw radiating lines from that point to the

points of division and to the extremities of the load line, and then, starting from A, or any other point in the vertical through that point of support, draw lines, successively parallel to the lines radiating from 0, and limited by the verticals through the weights, one such equilibrium polygon will be found.

This polygon was discussed in Part II., "Bridges," § 2. By moving the point 0 of the stress diagram, the place where the equilibrium polygon strikes the vertical drawn through B will be changed; and, if 0 is horizontally opposite the point which divides the load line into the two supporting forces, the polygon, drawn from A as a point of beginning, will strike B. But 0 may move on a horizontal line, and H will then have any value we please. H is therefore, at present, an unknown quantity; but we will suppose that A K I B is the desired equilibrium polygon for this given case, — an imaginary line, the ..ghts being attached to the arch.

3. **Relation between Equilibrium Polygon and Bending Moments.** — If the rib is made of a rigid material, the tendency to take a shape other than the one to which it was first formed will cause a bending action or moment at different points. Thus, between A and C the rib will flatten somewhat, moving towards the straight line A C, and from C to B it will become slightly more convex. At C, where the rib coincides with the equilibrium polygon, there will be no tendency to bend. The bending moments on either side of a point where the equilibrium polygon crosses the rib will therefore be of contrary kinds or signs. It is necessary to know the value of the bending moments at all points, in order to so design the cross-section of the rib that it shall be able to resist them. The point C is not necessarily the crown of the arch: it happens to come near it in our figure. If the arched rib is free to turn at its supporting points, no bending moments can exist there; if it is jointed or hinged at any place, as, for example, the middle or crown, no bending moment will be found there: the equilibrium polygon must therefore pass through all such points. The rib may be so fastened at A and B that it cannot turn in a

vertical plane; and there will then be bending moments at those points, as in the analogous case of a beam fixed at both ends, except for such a distribution of the load as makes the equilibrium polygon coincide with the arch at its ends.

If the rib is hinged at three points, that is, at the ends and middle, the equilibrium polygon is immediately fixed in position by the necessity of passing through these three points, and the problem of finding the stresses in the rib becomes very simple, as will be seen later.

4. **Value of Bending Moment.** — Let us suppose, at first, that the rib of Fig. 1 is jointed, and free to turn at its ends only. The stress diagram, 0 1 2, and the imaginary equilibrium polygon, having been constructed, and the horizontal line H from 0 drawn, it will be seen that this line will divide tL line into two forces, the vertical components of the abutı reactions, as proved in Part II., § 6. The arrows in the fig denote these components; and we will call the vertical ones, analogous to the supporting forces of a beam, P_1 and P_2, as marked. We have here the usual closed polygon of external forces.

Let an imaginary vertical section be made at D F: from the theorem of moments, as equilibrium exists in this loaded arch, the moments of all the external forces must balance around any point, for instance the point E, where the plane of section cuts the rib. If the sum of the moments around E equals zero, the moments on one side of the plane of section must equal those on the other; and, as E is in the section of the rib, these moments can only neutralize one another through the moment of resistance of the section: consequently, the sum of the moments on either side must equal the bending moment at E. Then at E, if P_2 and H are the rectangular components of the reaction at B, and Σ W. L denotes the sum of the products of each weight by its horizontal distance L from E, the bending moment will be

$$M = P_2 . DB - \Sigma W . L - H . DE. \quad (1.)$$

If the weights had been attached to the cord, or equilibrium polygon, we should have had, for moments on the right of and about F,

$$P_2 . DB - \Sigma W . L - H . DF. \quad (2.)$$

But a cord, being flexible, can resist no bending moment. As this cord is the equilibrium polygon, there can be no tendency to move or no bending moment at any point of it, and expression (2.) must reduce to zero, or

$$P_2 . DB - \Sigma W . L = H . DF.$$

Substitute this value in (1.), and it becomes

$$M = H . DF - H . DE = H . EF; \quad (3.)$$

which signifies that the bending moment at any point of an arched rib, under any vertical load, is equal to the product ' the vertical ordinate from that point to the *proper equilibrium polygon*, multiplied by H from the stress diagram.

5. **Remarks.** — It will be noticed that, to the left of C, D F − D E will change sign, becoming negative, and therefore that the bending moment will change in direction, as stated before. If the rib becomes straight and horizontal, the point E moves up to D, and the bending moment becomes equal to H . D F, which is its value for a beam supported at both ends.

The relation of the equilibrium polygon to the arch, or the fact that the bending moment equals H . E F, as just proved, may be readily explained in another way. Suppose that the arch A′ B′ of Fig. 14 has a single weight placed upon it in a certain position: it will thrust horizontally against the abutments an amount H. Let the equilibrium polygon for this weight, and having the same H, be A F B. The ordinates to this equilibrium polygon will be proportional to the bending moments due to the weight on a beam or truss of span A B, the moments will all be positive, and equal to H . D F. But the thrust H of the arch, which actually carries the weight, acting in the line A′ B′, will exert negative bending moments equal to H . D E at all sections of the arch. The resultant bending moment at any point, when the equilibrium polygon is superimposed on the arch, will be the product of H by the difference of these two ordinates, or H (D F — D E) = H . E F, at some places negative, and at others positive. Thus we see that, while we have for a given system of weights an equilibrium polygon exactly similar to those treated in Part II., "Bridges," the arch, by reason of its horizontal thrust which causes negative bending moments as above, annuls or cuts off a portion of the area of the equilibrium polygon, and the portion of the ordinate in excess or deficient at any point measures the existing bending moment. It is only necessary that the arch and polygon should have the same value of H. The arch, in its capacity of frame, as it were, carries a portion, more or less, of the forces which would otherwise cause bending moments and shears.

Such an arrangement of weights might be devised, continuously distributed along the rib, that there would be no tendency to change the shape of the arch at any point. The equilibrium polygon, becoming a curve for a continuous load, would then coincide with the centre line of the arch, and we should have what is termed an equilibrated rib. And, on the other hand, a rib can be designed for any given distribution of load, of such a shape as to be in equilibrium. This fact can sometimes be made use of when the load is definite, that is, not a moving load, and we shall refer to it again in the sequel.

6. **Condition to determine H; Invariability of Span.** — It may be noticed that in § 4 we used the term *proper equilibrium polygon.* It has been stated that it is easy to draw, between A and B, any number of funicular polygons, which have their angles on the verticals let fall from the weights, by simply moving the point 0 horizontally in the stress diagram, and thus altering the value of H, the horizontal component of the tension. But the actual rib, under a given system of weights, must have a fixed value of H, and definite bending moments at all points: there is therefore but one funicular polygon which will be the *proper* equilibrium polygon. Some condition must be imposed; and a sufficient one is, that, supposing the points A and B to be fixed in position relatively to one

another, the distance A B, or *the span of the rib, shall be unchanged.* An arch between two unyielding abutments satisfies this condition. If the curve A C is flattened by the pull upon it, or by the bending moments by which it is urged towards the straight line A C, the point C will move a little to the right, while the portion between C and B will become slightly more convex. The movement of the point B, however, with reference to A, must be zero.

7. **Formula for this Condition.** — Consider the arched rib as disconnected from its fixed points of support, but suspended in the air by the forces which were but now the reactions at those points. Equilibrium will still exist. The bending moment H . E F at E, from its effect on the particles at that section, causing an elongation of the fibres on one side and a compression of the fibres on the other side, produces what may be called an exceedingly small angle in the rib, or, better, a *change of inclination*, at E, moving the free end B, so far as this change alone is concerned, a very small distance in a direction perpendicular to a straight line from E to B. The amount of this displacement will depend upon the distance E B, and upon the change of inclination at E, which change has just been shown to depend upon the bending moment H . E F. The amount, B R, of this movement, is greatly exaggerated in the figure. But the horizontal component, or projection, B S, of the displacement, which alone affects the horizontal distance of B from A, will manifestly, from the proportionality of the sides of the two right-angled triangles B R S and E B D, be to B R as D E is to E B, or B S will be proportional to D E.

Perhaps this point may be brought out more plainly if stated algebraically, thus: —

$$BR \text{ varies as } EB \cdot H \cdot EF;$$

$$BS = BR \cdot \frac{DE}{EB}; \text{ therefore,}$$

$$BS \text{ varies as } \frac{EB \cdot H \cdot EF \cdot DE}{EB}, \text{ or as } H \cdot EF \cdot DE.$$

Taking all the points in the rib into consideration, we see

that the total horizontal displacement of B from A will be proportional to H . Σ E F . D E, if Σ is the sign of summation of all of the products E F . D E. As the span A B is to be unchanged, the above quantity must equal zero, and therefore, as H cannot be zero, we have the desired condition reduced to

$$\Sigma \text{E F} . \text{D E} = 0. \quad (1.)$$

8. **The Equilibrium Polygon determinate.** — As E F changes sign when the equilibrium polygon crosses the rib, as at C, we arrive at this result for a rib free to turn, or hinged, at its ends, that the *summation of the products* E F . D E *for every point where the equilibrium polygon lies on one side of the rib must equal the summation of the similar products for every point where the polygon lies on the other side.* Only one polygon, manifestly, will satisfy this condition; for, if we draw a new polygon between A and B, we immediately increase one set of E F's and diminish the other. An equilibrium polygon may first be drawn tentatively, ordinates be measured at intervals, and the above products computed. It will then be readily seen whether the polygon should be moved up or down; to move it, change H, and draw again. We can deal thus with a rib of any outline; but, for the regular forms of arches commonly in use, we will show presently how to determine the exact equilibrium polygon without experimental trial.

9. **Deflection of the Rib.** — The vertical component R S, of the displacement B R, manifests itself, since B cannot move, by a slight movement of the rib at E vertically, corresponding to the deflection of a beam under transverse forces.

10. **Another Value for Bending Moment.** — It has been shown that the bending moment at E equals H . E F. If we draw from E a perpendicular, E N, to that side of the equilibrium polygon which passes through F, the side being prolonged if necessary, we shall form a right-angled triangle, similar to one formed in the stress diagram by H, the line parallel to the side of the polygon, and the vertical line. Thus, in Fig. 1, the triangle E F N will be similar to 0 2 5, and we may write the proportion

$$0\text{-}2 : 0\text{-}5 = EF : EN;$$

or, if T denotes the tension 0–2 in the part of the cord which passes though F, we get, upon multiplying extremes and means,

$$H \cdot EF = T \cdot EN; \quad (1.)$$

so that the bending moment at each point is also equal to the product of the tension in the cord by the perpendicular let fall on the cord from the given point; and this is the measure of a *moment*, as shown in mechanics. The discussion of the bending moment might have been approached in this way.

11. **Combined Effect of Bending Moment and Direct Force.**—If a force T acts in the line A K, which, when we consider the curved rib, is an imaginary line, its moment with respect to the rib at E is, then, T . E N. Now, from mechanics, if we analyze the effect of a force T, Fig. 2, at any distance laterally from a point E, we may apply two equal and opposite forces, +T and −T, at this point, which is here the middle of the rib, or what would be, for flexure only, the neutral axis, without destroying the equilibrium. Hence we have at E the direct force +T, producing tension, and the couple T . E N, producing flexure. The enlarged sketches will represent the condition of the rib. The small arrows at E′ denote the magnitude or intensities of the stresses which form the moment of resistance to balance the bending moment, these intensities being taken as uniformly varying, a supposition which is satisfied within the elastic limit; at E″ are shown the stresses on the particles of the section from the direct force; and the combination of the moment and force is represented at E‴, it being understood that these several views represent one and the same section E.

The point of no stress, or the position of the neutral axis, is seen to be shifted from the middle of the section at E′ to one side at E‴; and it will disappear altogether when the arm of the couple or moment becomes sufficiently small, so that the entire section may be under one kind of stress of varying intensity. If we know the form of cross-section of the rib, we can tell from the location of the equilibrium polygon, by simple inspection, where we shall find both tension and compression, and where only one kind of stress. This matter will be touched upon later: §§ 106–108.

12. **Reversal of Figure; Movement of Rib from Equilibrium Polygon.** — When an arch is under analysis, the figures thus far given will be inverted. Imagine them to be so. All of the forces will then be *reversed.* The polygon which was under tension will be compressed, and its sides will represent struts. It will be in unstable equilibrium, and its relation to vertical forces is not, perhaps, so readily apprehended, by one not acquainted with this subject, as is that of the funicular polygon. For this reason it was thought best to take an inverted arch first. Hereafter the arches will be drawn above the springing line; H becomes the *horizontal thrust* of the rib against its abutments.

The curved rib, between the points A and C, Fig. 1, so long as there is tension along the straight line A C, tends to move towards that line, just as the cord, if drawn towards the arch, returns to its former position; but as soon as the figure is inverted, and C is forced by compression towards A, the arch tends to *move away from the equilibrium polygon.* This fact is true of all points of the rib, and, being borne in mind, will enable one to tell at a glance the kind of moment at each point of the rib. All the bending moments are therefore reversed. Those bending moments which tend to make the arch flatter, or of less curvature, at any point, are called positive; those which tend to make it more convex are called negative.

It may aid in fixing the ideas, to take a piece of small steel wire, bend it into the arc of a circle, and, placing the two ends in two notches upon a board, notice the change of shape arising from a pressure or load imposed on any portion. The movement of the wire will indicate, in a general way, where the equilibrium curve lies in reference to the rib.

13. **Equilibrium Polygon for a Single Load.** — It is now readily seen that the equilibrium polygon for a single, concen-

trated load on an arch is composed of two straight lines which meet on the vertical drawn through the point where the load is imposed. In the case just treated, these lines will start from the two springing points of the arch. The only quantity needful to fix their position will be the distance of their point of intersection vertically from the rib; and the single condition of (1.) § 7, that Σ E F . D E = 0, will determine the unknown quantity. It will be easier to find the effect of a single load at successive points on the arch, and to combine these effects for any possible arrangements and intensities of load, than to treat at once several loads. We shall pursue this method.

Direct Force and Shear at a Right Section. — Since the arched rib is often composed of two flanges, and a web or connecting bracing, similar to a girder or truss, we desire, after we have found the bending moments at all points, to find that portion of the vertical force or the shear at each section which must be resisted by the web members. *Shear* was explained in Part II., "Bridges," § 4. In a horizontal beam, carried on two supports, we should have, in Fig. 1, P_2 for the supporting force, and shear on the right of any section between B and W_1; $P_2 - W_1$, or (1–5) — (3–1), for the shear anywhere between W_1 and W_2; $P_2 - W_1 - W_2$, or (3–5) — (4–3), that is — (5–4), between W_2 and W_3; and so on, subtracting each weight from the previous shear or resultant. But in a beam, or a truss with horizontal chords, the other forces, those which oppose the bending moment, are horizontal: here they are not. Supposing the rib to be inverted, the direct thrust, being in the direction of a tangent at the centre line of the rib, has a vertical component which affects the amount of shear to be resisted by the web. In short, the inclined flanges or chords act as braces; and we have, at any section, these chords as well as the web members, among which to distribute the shearing force. The action corresponds with that of the bow in a bowstring girder.

It is not probable that the thrust in the side of the equilibrium polygon will be parallel to the tangent to the curve of the centre line of the rib at a particular section, but this thrust will be the resultant force at the section. It may then properly be resolved into two rectangular components, one perpendicular to the section, representing the direct force, and the other parallel to the plane of the section, representing the shear. The direct stress, combined with the tension and compression due to bending moment, will be resisted by the flanges or chords, and the shear by the web members, if the rib is so constructed. If the rib is of solid section, like a beam, the separate consideration of shear is generally unnecessary. It will at once be seen that the direct stress at any point of the rib is obtained by projecting the force in that side of the equilibrium polygon which passes near the point upon the tangent to the rib. Thus, in Fig. 1, 0–3 is the tensile force in the side I G of the equilibrium polygon, and 0–6 is drawn parallel to the tangent at U: if a perpendicular were drawn from 3 upon 0–6 prolonged, the distance from 0 to the foot of the perpendicular would be the direct stress, and the perpendicular itself would be the shear on a right section at U. Or, again, if 0–2 is the force in A K, and 0–7 is parallel to the tangent at Q, a perpendicular from 2 on 0–7 will cut off the direct stress, and be itself the shear at Q.

15. **Sign of Shear; Maximum Bending Moment at Point of Zero Shear.** — The above points may be made more clear, if necessary, by reference to the sketch above and on the left of Fig. 8. Let A C represent a portion of an arch, and A R′ a portion of the equilibrium polygon which exerts a thrust R at A. The components of the abutment reaction will be H, the horizontal thrust, and P_1, the vertical force. But R may also be decomposed, on a right section of the rib *near* A, into T direct thrust and F shear at the section. The little sketch adjoining shows, that, as these components act on the left of the section, we must have the opposite shear on the right of the section, giving what we have been accustomed to call negative shear (see Part II., "Bridges"). When, at any right section, a line parallel to the side of the equilibrium polygon lies above the tangent to the rib, the forces being taken on the

left of the section, as is the case at C, where T′ and F′ are the components of R′, the shear will be positive. Where the side of the equilibrium polygon is parallel to the tangent to the rib, as for instance near *d*, at that point there will be no shear, and the shear will be of opposite signs on each side of such point. The direct stress there will be H multiplied by the secant of the inclination of the tangent to the horizon.

As the maximum ordinate between the side of the equilibrium polygon and the arch occurs where the side of the polygon is parallel to the rib, the maximum bending moments in the arch, as in a beam or truss, are found at points of no shear.

16. **Treatment of Arch with Fixed Ends requires Three Conditions.**—If the arched rib is fixed in direction at the ends (in place of being free to turn as previously supposed), by being firmly bolted to the abutments, or by having square ends accurately bedded upon the skewbacks, a bending moment will generally exist at the points of support when the arch is loaded. By taking the piece of easily flexible wire before mentioned, clamping the ends firmly, so as to fix the wire in the position of an arch, and then applying a load or the pressure of the finger, one can easily verify this statement for himself; and he will see that, for many positions of the load, the bending moment at one abutment is of the opposite kind to that at the other. The points at which the equilibrium polygon begins and ends will no longer be A and B of Fig. 1, and some new conditions must be imposed in order to determine these points.

Consider the effect of a single load upon the arched rib A C B of Fig. 3, which rib is fixed in direction at its ends. The equilibrium polygon will be two straight lines, such as I N and N L; and, as there may be bending at both points of support, it will be necessary to find the magnitudes of A I and B L, as well as of N G, three unknown quantities. Three conditions must therefore be satisfied. Such writers as, in treating the arch either graphically or mathematically, require but two conditions to be fulfilled for an arch with fixed ends, err in their assumptions, and hence in their results. If two conditions only are imposed, where three are necessary, many polygons can be drawn, and the problem is left undetermined.

17. **First Condition.** — One condition which must be satisfied is plainly the one already used, §§ 6 and 7, that the change of span A B shall equal zero, or that

$$\Sigma \text{ EF.DE} = 0.$$

18. **Second Condition: Change of Inclination between Abutments equals Zero.** — As the *change of inclination* between any two contiguous points is directly proportional, in direction and magnitude, to the bending moment (for the elongation and compression of the fibres on the two sides, upper and lower, of the rib, result from this bending moment, and cause whatever change of direction or inclination the rib may take on), and as the bending moment has been proved to be proportional simply to the ordinate E F, the change of inclination at any point is proportional to the ordinate E F from that point of the rib to the equilibrium polygon.

The reader must distinguish between the change of inclination produced by flexure, and the original inclination of the rib to the horizon at each point due to the curve to which the rib is constructed. If an arch is loaded, it assumes a form slightly different from its shape when unloaded. The angle, at any particular point, between the two tangents to the curve of the rib, before and after it is loaded, is the *change of inclination* at that point.

Starting from A, Fig. 3, the total change of inclination at C will be proportional to the sum of all the ordinates between A and C. On the other side of C, where the straight line crosses the rib, the bending moment being of the opposite kind, the changes of inclination will be in the opposite direction, and, in any summation of ordinates, for instance from A to E, must be subtracted. Then, as both A and B are fixed in their original directions, if we sum up all of the ordinates E F, from A to B, the total change of inclination between abutments is zero, and this sum must be zero. Therefore the second condition to be realized is that

$$\Sigma\, E F = 0;$$

or that *the sum of all the ordinates between the arch and the equilibrium polygon on the inside of the arch must equal the similar sum outside.*

19. **Third Condition: Deflection between Abutments equals Zero.** — Fig. 3 shows that, since the displacement B R of B, relatively to the point E, in case B could move, has been proved, by § 7, to be proportional to H . E F . E B, the vertical component of this displacement varies as H . E F . D B; for, by a similar proportion to the one used in that section,

$$S R = B R \frac{D B}{E B}; \text{ therefore,}$$

$$S R \text{ varies as } \frac{E B \,.\, H \,.\, E F \,.\, D B}{E B}, \text{ or as } H \,.\, E F \,.\, D B.$$

If the products E F . D B should be summed up for all points from A to Q, for example, we should get a quantity proportional to the vertical displacement of Q, arising from the separate minute displacements between A and Q. If we pass beyond C, we have products of an opposite sign; and it then appears, that, since the ends at A and B are fixed both in position and direction, the sum of all the products between A and B must equal zero, or, since H cannot equal zero,

$$\Sigma\, E F \,.\, D B = 0. \quad (1.)$$

Therefore the third and last condition is, that the *sum of the products of each ordinate, between the arch and the equilibrium polygon on the inside of the arch, by its distance from one springing point, must equal the similar sum on the outside.* It is immaterial which springing is chosen, but all the D B's must be measured to the same abutment.

20. **This Condition not applicable to Hinged Rib.** — It may be expedient to dwell upon this equation a little longer; for the question will apparently arise, why this condition is not also properly applicable to an arch which is jointed or hinged at

the ends. Let a tangent A K be drawn to the rib at the point A, and a vertical line be dropped from it to the point Q. If the arch is now bent at the point E′, by a bending moment which is proportional to E′ F, the point Q is moved a distance proportional to E′ F multiplied by the distance from E′ to Q; but the distance which Q moves in the vertical line Q K will be proportional to E′ F multiplied by the horizontal projection of E′ Q, or D T, and similarly for moments at all other points between A and Q. As the tangent at A is fixed in direction in this case, the movement of Q away from the extremity of K Q, or its movement in relation to the tangent at A, will be proportional to the summation of the E F's multiplied by the D T's; and as the abutment B is fixed, the distance of B from a tangent at A must be unchanged by any load, or its displacement must be zero, as above. In the case of the rib hinged at the ends, while the above area moments give the deflection from the tangent at A, this tangent is not fixed, but changes in direction upon the imposition of a load, and this condition cannot be applied. If, however, one should treat an arch which was fixed at A and hinged at B, this condition would be necessary, and all the distances D B would be measured to the hinged end; while the second condition would not apply, and would not be needed.

This third condition was first applied to the determination of the bending moments in continuous bridges and pivot draw spans, in the first edition of Part II. of this work.

21. **Remarks: Abutment Reactions; Shear,** &c. — The arch of Fig. 3 is cut by the equilibrium polygon in three places, and it may be cut in four points, giving as many places of contraflexure. The areas on opposite sides of the rib represent bending moments of opposite kinds, and of which kind is readily known if one remembers that the arch under thrust always moves from the equilibrium polygon. The amount of the weight, not being contained in any of the equations of condition, does not affect the diagram; for H is constant for all points of the arch for any given vertical load, and, not being

equal to zero, is thrown out of the equations. But the weight W does affect the value of H.

If 1–2 represents W in the stress diagram of Fig. 3, and 1–0 and 2–0 are drawn parallel to N I and N L, 0–3 drawn horizontally will determine the horizontal thrust H, while the load-line will be divided at 3 into the two vertical components P_1 and P_2 of the reactions as marked. These vertical forces are not the same as would be obtained for the case previously considered, nor for a beam only supported at the ends. Such forces would be equal to the divisions of 1–2 made by a line drawn through 0, parallel to a line from I to L. If we notice the arrows drawn at the abutment A, we see that, supposing P_1 were at first the fraction of W due to the position of G, or $\frac{GB}{AB}W$, we have also at A, besides the horizontal thrust H, a couple H . A I. There is another couple at the other abutment, which may be of the same or opposite kind; their algebraic sum can only be balanced by vertical forces at the two abutments acting with a lever arm of the span; and these vertical forces must be added to one reaction, and subtracted from the other, bringing P_1 and P_2 to the amounts found by the stress diagram. The effect of the couple is the same as if P_1 had been calculated for the point where N I would meet the horizontal line. This is another example of the principle in mechanics cited in § 11.

The remarks on shear in §§ 14, 15, apply equally well here. The direct compression in the rib at any point is obtained, as before, by drawing a line through 0 parallel to the tangent to the rib at the point in question, and dropping a perpendicular upon it from the extremity of the line which represents the stress in the adjacent side of the equilibrium polygon. Thus the compression at E will be the distance from 0 along 0–4 produced to the foot of a perpendicular from 2. Recalling the three conditions just stated, it will be evident, that, while it will be possible to adjust the two lines of the equilibrium polygon to their proper position by successive trials, it will not, as in the former case, be easy. The three ordinates, A I, G N, and B L, can, however, be computed quite readily, and the remainder of the process is very simple. The statements so far made apply to a structure of any outline, so long as it acts as an arch, although some modification will be called for when the cross-section and the depth vary very much, or when what is known as the moment of inertia is not practically constant; but, for forms other than regular curves, the application of these conditions must probably be made by trial.

21*a*. **Shear at a Vertical Section.**—The relation of the equilibrium polygon to the arch which was pointed out in § 5, Fig. 14, shows how the shear at any vertical section of a loaded rib is affected by the curvature of the arch. In the same way that the ordinates of the rib may be superimposed on those of the triangle which represents the equilibrium polygon for a single load, the two shear diagrams may be placed on one another. One will have the form of *a i m n l*, Fig. 8, conforming to the load which gives the curve of Fig. 14, and found from the amount of vertical reaction which, combined with H, will give a direct thrust at the springing; the other will resemble *a d e f g l*, Fig. 8, the usual shear diagram for a single load, which load produces the triangle of Fig. 14. The flanges of the arch take up at each point an amount equal to the ordinates from *a l* to *i n*, and the web or bracing carries the remainder, which will be positive at some points and negative at others, as marked in the Figure. Thus we see that, through the direct thrust, the arch is relieved of a portion of the *truss stresses* due to both bending moment and shear.

CHAPTER II.

ARCH HINGED AT THREE POINTS.

22. **Three-hinged Arch.** — Before taking up for treatment any arches of special curves, we will notice the simple case of a rib, of any form, hinged at both ends and the middle, or, as it is sometimes called, the "three-hinged arch." The three hinges or joints may be located anywhere, and two of them may be placed near together at one abutment, reducing the portion of arch between them to a short link or strut, which necessarily lies in the direction of the thrust at that abutment. For the ribs of this chapter it has been stated that the equilibrium polygon or curve is at once definitely located. If a single load is placed at K, on the arch A D B of Fig. 4, hinged at A, D, and B, one of the two straight lines composing the polygon must, starting from A, pass through D, while the other, starting from B, must meet the former on the vertical line drawn through K, as required by the principle of the funicular polygon: A C B, therefore, is the polygon. If 2–1 represents the weight at K, and 2–0 and 1–0 are drawn parallel to C B and A C, 0–3, drawn horizontally, will give the horizontal thrust, while 1–3 and 3–2 will be the vertical components of the reactions at A and B. Let it be remembered that the total reaction of the abutment at A is, and is in the direction of, 1–0, although it is often convenient to decompose it into P_1 and H.

A load vertically below E will, similarly, have for its equilibrium polygon A E B. For different positions of the weight between D and B, all of the vertices of the polygons will be found on the straight line D L, and the portion A D does not change for any movement of the weight on the right half of the arch. A weight on the left half will simply reverse the diagram. The dotted lines show the equilibrium polygons for a weight at such successive points as divide the half-span into five equal horizontal parts, and the corresponding changes in the value of H will be seen in the stress diagram on the left.

23. **Formula for H.**—If F D, the height or rise of the arch, is denoted by k, the half-span A F, $=$ F B, by c, and the horizontal distance F G, from the weight to the middle of the span, by b, we shall have $A\,G = c + b$, and $G\,B = c - b$. From the similarity of triangles A D F and 0 1 3, we may write,

$$3\text{–}0 : 3\text{–}1 = c : k, \quad \text{or} \quad H : P_1 = c : k.$$

By the usual rule,

$$P_1 = \frac{c - b}{2\,c} W;$$

therefore

$$H = \frac{c - b}{2\,k} W.$$

The quantity $c - b$ is to be understood to mean the horizontal distance from the weight to the nearer abutment. H is seen to decrease regularly as the weight moves from the middle of the span.

24. **Stone Arches.**—In the treatment of stone arches it has often been assumed by writers that the equilibrium curve passed through either the middle of the depth of the keystone or some other arbitrary point within the middle third of its depth; and a similar assumption would then be made for the springing-points. Such a treatment immediately reduces the stone arch to this case, and the equilibrium curve can at once be drawn. As such an assumption does not seem to be warranted, it is not thought expedient to go into the case of the stone arch until later (Chap. IX.); but the reader who desires to look up such a mode of handling the problem is referred to a paper by William Bell, in the Transactions of the Institute of Civil Engineers of

Great Britain, vol. xxxiii., reprinted in Van Nostrand's "Engineering Magazine," vol. viii., March to May, 1873.

25. **Example.**— We will, as an example, show how to draw an equilibrium curve for an arch which is loaded uniformly along its rib. Such a distribution will conform quite well to that of the steady load on an arched roof. For definiteness, let the pointed arch of Fig. 5 be of 80 feet span, 40 feet rise, the two arcs having a radius of 60 feet, and let it be loaded with 500 pounds per foot of the rib. We may, if we please, divide the rib into a convenient number of equal portions, which divisions will give us a number of equal weights to be laid off on the load line. Otherwise we may space off a number of equal horizontal distances. In either case, the load of each space will be considered as concentrated at its centre of gravity; and, if the spaces are small enough, the centre of gravity may, without sensible error, be taken as coinciding with the middle of each space. For the sake of reducing the number of lines, so as to avoid confusion in a small figure, we have divided the half-span into four parts, of ten feet each, measured horizontally; and their centres of gravity will be assumed to be at five feet, fifteen feet, &c., from the point of support. Draw verticals through these centres of gravity, D, E, F, and G.

To find the weight on each division: The lengths of the several portions of arc may, with sufficient exactness, be considered the same as the lengths of their chords, which chords are perpendicular to the radii which pass through D, E, &c. If, then, the load on ten feet is 5,000 lbs., draw $a\,b$ horizontally and equal, by any scale, to this amount; then will $b\,g$, $b\,f$, $b\,e$, and $b\,d$, drawn parallel to the respective chords, give the amount of load on each division, at the successive points G, F, E, &c. Upon scaling these amounts we will lay them off upon a vertical line, from 1 to 5. In order to cause the equilibrium polygon to separate from the rib sufficiently to be easily seen in this small figure, we have taken the liberty of doubling the load on D, thus making it 4–6, in place of 4–5. The loads will therefore be, successively, about 5,400 lbs., 5,900 lbs., 7,000 lbs., and

$2 \times 10{,}000$ lbs., or 20,000 lbs., from G to D, and from 1 to 6. Since $H = \Sigma \frac{c-b}{2k} W$, we have for its value

$$\tfrac{1}{2} H = \frac{35 \times 5{,}400 + 25 \times 5{,}900 + 15 \times 7{,}000 + 5 \times 20{,}000}{80} = 6{,}769 \text{ lbs.}$$

If the given load were unsymmetrical with regard to a vertical through C, it would be necessary to calculate the two vertical components of the reactions at A and B, or P_1 and P_2, the reaction at B being laid off from that end of the load line from which was measured the load nearest to B, and then to draw a horizontal line from the point of division between P_1 and P_2, on which to lay off the value of H. But, if both sides of the roof are loaded alike, half a diagram and half an equilibrium polygon will be sufficient. The load on the half-arch being 1–6, 6–1 will be the vertical component of the reaction at B, and H will be laid off in the direction 1–0. Since we have calculated H for only one-half of the entire load, the above quantity must be doubled, and the total horizontal thrust will be 13,538 lbs., = 1–0. The reaction at B is therefore 6–0.

Nothing remains but to draw, first a line from B to the vertical through D, parallel to 6–0, then one, parallel to 4–0, from the end of the last line to the vertical through E, and so on, the last line, parallel to 1–0, passing through the hinge at C, as required. The polygon on the side C A will be exactly similar. It is well to have the points of division quite numerous. The maximum ordinate between the rib and the equilibrium polygon, multiplied by H, gives the maximum bending moment.

26. **Caution.** — As this is the first example, it may be well to pause here, and renew the caution to the draughtsman to lay off the polygon of external forces in the order in which the forces are found in going round the arch or truss; otherwise he will fail to make his equilibrium polygon close on the desired point. Thus, beginning at G, he should have the weights at G, F, E, &c., or 1–2, 2–3, 3–4, &c., plotted, one after the other, down the vertical load line in the direction of their action, until the point

B is reached, for which he draws 6–0, from 6 to 0. Then the point A gives a similar line from 0, slanting upwards toward the right; and the remaining loads on the left half of the arch come down a vertical line, and close on 1, the starting-point. The decomposition of 6–0 into 6–1 and 1–0 does not alter the case. If we had gone round the arch in the opposite direction, this stress diagram would have been reversed, or turned 180°.

27. **Relation between Equilibrium Polygon and Curve.** — The true equilibrium curve, for the load uniformly distributed along the rib, is a curve which will be *tangent* to the sides of the funicular or equilibrium polygon just drawn. The closer together the points D, E, &c., are taken, the nearer the two will come together. If the points at which the loads are concentrated divide the span into equal portions, that is, if the end distances are the same as the others, so that the portions of load near B and C are concentrated on those points, or, even with unequal spacing, when the load between each two assumed points is carried by those points as required by the principle of the lever, the true equilibrium curve will pass through the *vertices* of the equilibrium polygon. Such a distribution of load is made in roofs and bridge trusses, when a half panel weight is thrown on each abutment. Compare Part II., "Bridges," § 58.

The curve assumed by a rope or chain, of uniform weight per foot, when suspended between two points, is called a *catenary*. Since the equilibrium curve in Fig. 5, if we had not placed the extra weight on D, would have come quite near to the rib, it would have been a close approximation to a catenary. As we expect to make some use of this curve later, we will show how to draw one at that time.

28. **The Parabola the Equilibrium Curve for a Load Uniform horizontally.** — If the load on this arch were distributed uniformly horizontally, the curve of equilibrium would be a parabola. In case the whole arch were a parabola, with the vertex at the crown, and the load extended over the entire span, the two curves, coinciding at the springing-points and crown,

would be identical throughout, and the rib itself would be in perfect equilibrium. This same point was brought out in reference to the parabolic girder, Part II., "Bridges," § 73. That the parabola is the equilibrium curve for a continuous load, distributed uniformly horizontally, may be shown as follows: —

Let A B, Fig. 6, be a portion of a cord, horizontal at A, which is in equilibrium under such a uniform load, represented by A C, suspended from the cord. The tension at A will be in the line of the tangent A C; the resultant of the load A C will be vertical, and must pass through its middle point D. As the cord A B is in *equilibrium* under its load and the reactions or tensions of the other portions of the cord at A and B, the tension along the tangent at B must, by the principle of the triangle of forces, also pass through D. As B C, drawn vertically, is parallel to the resultant of the load, the sides of the triangle B C D will be proportional to the three external forces; and, if A C $= x$, B C $= y$, W = total load on A B, $= wx$ (where $w =$ load per unit of length), and H = tension at A, we have

$$W : H = BC : DC = y : \tfrac{1}{2}x,$$

or

$$y = \frac{Wx}{2H} = \frac{w}{2H}x^2,$$

the equation of a parabola with vertex at A.

Therefore an arched rib of parabolic form, when loaded uniformly horizontally, has no tendency to change its shape, that is, experiences no bending moment, at any point.

29. **Suspension Bridge.** — A B of Fig. 6 may represent a suspension bridge cable, A C being the half-span, and C B the height of the tower: hence, if A C $= c$ and C B $= k$, we have for the tension in the cable at the mid-span, § 28,

$$H = \frac{wx^2}{2y} = \frac{wc^2}{2k}.$$

The tension T at the tower will then be proportioned to H, as B D to D C, or as $\sqrt{k^2 + \frac{1}{4}c^2}$ to $\frac{1}{2}c$; therefore

$$T = \frac{wc}{2k}\sqrt{4k^2 + c^2}.$$

Each suspending rod must carry the greatest weight that can come at its foot. The pressure on the top of the tower from the half-span will be the weight of the half-span, or *w c;* to this must be added the vertical component of the tension on the anchorage side of the tower. If the cable has the same inclination *both ways*, at the top of the tower, the pressure is 2 *w c*.

The manner of stiffening a suspension bridge to resist the tendency to distortion under a partial load is treated in Chap. X.

30. **Equilibrium Curve for Partial Load.** — If the load extends over a portion only of the span of the arch, and is uniformly distributed horizontally, the curve for the loaded portion is parabolic, while that for an unloaded portion is a straight line: thus, if the load extends from one abutment to the middle, we shall have, on the unloaded half, a straight line from the abutment to the crown, and, on the loaded half, a parabola from the crown to the springing. As it was proved in Part II., "Bridges," § 10, that any two sides of the funicular polygon, when prolonged, meet on the vertical drawn through the centre of gravity of so much of the weight as is included between these sides, the equilibrium curves for any cases where the rib is hinged at three points can be drawn without previously determining the value of H. Thus, in the case just supposed, of a load over the half-span, from B to F in Fig. 4, the centre of gravity will be at G. Then, if G C is the vertical drawn from G, the side of the funicular polygon, or, more properly, the tangent to the equilibrium curve, at B, must pass through C, where C G meets A D, and the required parabola will be drawn from D to B on D C and B C as tangents. As one point of the curve we have the middle point of a line from C to the middle of the chord D B. We can then find H by drawing 1–0 and 2–0, parallel to A C and C B. Henck's "Field Book for Railroad Engineers" gives methods for constructing parabolas; two constructions are given in Part II., "Bridges," §§ 20 and 28, one of them applying when two tangents are given.

31. **Suggested Examples.** — We would suggest the following examples for practice: 1st, Given a semicircular rib, loaded

uniformly horizontally over the whole span, and pivoted at the crown and springings: find that the maximum bending moment occurs at 30° from the springing, and is equal to one-sixteenth of the total load multiplied by the radius of the arch, while H is equal to one-fourth of the total load. 2d, Given a parabolic arch similarly pivoted, and in equilibrium under a steady load distributed as above; add a similar travelling load from one abutment to the middle of the span: prove that the maximum bending moment is found at one-fourth of the span from either abutment, is of opposite signs at these two places, and is equal to one thirty-second of the travelling load then on the arch multiplied by the span, while H for the travelling load equals the same product divided by one-fourth the rise of the arch, and for the steady load is twice as much.

32. **Extent of Load to produce Maximum Bend. Moment.** — It may be desired, when designing an arch of this type, to find the extent of load which will produce the maximum bending moment at each point, and the value of that moment. Suppose the point N, Fig. 4, to be examined: prolong B N until it meets A D at E; it is then manifest that any load in the vertical through E will cause no bending moment at N; that the equilibrium polygon for any load on the right of E will pass outside of the arch at N, while the equilibrium polygon for any load to the left of E will pass inside of N. Therefore the maximum bending moment at N of one kind will be found when all possible loads are put on the arch from B to the vertical through E, and the maximum moment of the other kind occurs when the load extends from A to E. As the arch tends to move away from the equilibrium polygon, the kind of moment is easily distinguished. H can then be found, the equilibrium curve drawn, the ordinate scaled and multiplied by H.

33. **Braced Arch.** — For the reason that the equilibrium curve is at once definitely located by introducing three hinges or pivots, no matter what form the arch may have, that type which used to be known as the braced arch, having a horizontal

upper and a curved lower member, the spandrel being filled with bracing, has usually been treated as free to turn at both crown and springings; in that case a diagram may be drawn by Clerk Maxwell's method, as set forth in Part I., "Roofs," or the stresses may be found from the equilibrium curve. A braced arch, hinged at crown and springings, with an elliptical lower and a straight upper member, carries a track of the Pennsylvania Railroad over Thirtieth Street, Philadelphia. (See "Engineering," July 22, 1870.) While a diagram only gives the stresses in the various members for one position of load at a time, one can determine all the maximum stresses by two diagrams and a tabulation, not difficult to one familiar with such methods. The way to be pursued will be found in Du Bois' "Graphical Statics," appendix, § 7, p. 350. We will explain another treatment in Chap. XII.

34. **Shear; Temperature.** — Since it is not practicable to draw a shear diagram until the form of the rib is defined, we can only, at present, refer the reader to § 14. After we have discussed the parabolic and circular ribs, the reader can doubtless work up any special design of the present class for himself.

One advantage possessed by this type of arch is that changes of temperature have no straining effect, for the crown rises and falls without affecting the two halves of the arch injuriously. If the crown sinks a little, the value of H will be seen from Fig. 4 to be very slightly increased, while the equilibrium polygon will practically go with the arch.

CHAPTER III.

INTRODUCTORY TO PARABOLIC ARCHES.

35. **Parabolic Arch.** — We propose to apply the facts which have been developed thus far to the arch whose centre line is a parabola. This curve is chosen as one form; because it is, as proved in § 28, in perfect equilibrium under a load distributed uniformly horizontally over the entire span. As in the case of a suspension bridge, so in some arches of iron, most of the steady load consists of a platform and such other parts as are distributed in accordance with this requirement (the arch itself and the vertical posts which carry the platform giving a somewhat greater intensity per horizontal foot as we approach the springings), so that, for the former portion, as well as for the travelling load over the whole span, the arch will be subjected to no bending moments, and no shear; hence there will be no stress in the bracing. Then, again, the parabola for a given rise and span is easily plotted and designed; and, lastly, the determination of the equilibrium curves, for the cases taken up, will be simpler than for circular arcs, and will naturally prepare the way by rendering the reader familiar with the steps of the analysis. It may be well to add here that a circular segmental rib, whose rise is not more than one-tenth of its span, is so nearly coincident with a parabolic arch of the same span and rise, that the investigations which follow will apply with sufficient accuracy to such flat segmental ribs.

36. **Vertical Deflection of an Inclined Beam.** — Let us

consider the two cases of a horizontal beam and of one inclined to the horizon at an angle *i;* it is known from the usual formulæ for deflection, Part II., "Bridges," Chap. VI., that, other things being equal, the deflection of a beam is directly proportional to the load and the cube of the length. If, then, the inclined beam is of a length l, and the horizontal one of a length $l \cos i$, as shown in Fig. 7, the deflection of each, measured *perpendicularly* to the respective beams, will, as regards length only, be in the ratio of l^3 to $l^3 \cos^3 i$. But, if each carries the same load W, the *transverse* component of W, which alone causes flexure of the inclined beam, the longitudinal component producing direct compression, will be W $\cos i$; whence the deflection perpendicular to each beam will, for similar points, be proportioned as 1 to $\cos^2 i$. And, again, the vertical component of the deflection of the inclined beam will be to the perpendicular amount as $\cos i$ to 1; whence the *vertical* deflection of the inclined beam will be to that of the horizontal beam of the same cross-section as 1 to $\cos i$. As the stiffness of a beam is directly proportioned to its breadth, should the inclined beam be made broader in its horizontal dimension than is the horizontal beam, in the ratio of 1 to $\cos i$, the depth being unchanged, the *vertical* deflections of the two beams for the same load would be exactly the same.

37. **Application to Arches.** — Any very small portion of an arch, taken within such narrow limits as to be considered straight, behaves like the inclined beam, as regards its flexure under a load; and therefore it follows, that if an arch has the dimension perpendicular to its face increased, from the crown to the springing, in the ratio of the secant of the inclination to the horizon, it may be discussed as if it were a beam of uniform cross-section, of the same span, similarly supported, and carrying the same load which produces flexure. In the arch some of the load does not produce flexure; in the parabolic rib, for instance, before cited, a uniform horizontal load gives equilibrium. We propose, in our analysis of the parabolic rib, to make this supposition, that the rib is broader at

the abutments than at the crown in the ratio just mentioned, and thus to simplify the work of investigation. Iron arches whose flanges or chords are thicker, as we approach the springing, in the above ratio, while the perpendicular depth between the two flanges is constant, practically satisfy this case. In this class of ribs the intensity of the direct thrust on the square inch for a complete uniform load will be the same at all cross-sections.

As we desire the reader to reproduce, on a much larger scale, the figures and problems for himself, we remind him that points on the curve of a parabolic rib are easily found by the construction of Fig. 8, Part II., "Bridges."

PARABOLIC RIB, HINGED AT ENDS.

38. **Equilibrium Polygon for Single Load.** — Taking the case of the parabolic rib, hinged at the ends only, let us place a single weight at the point I, Fig. 8. If the lines A C B fulfil the condition of § 7, that the sum of the products of the ordinates D E and E F for all points of the arch equals zero or

$$\Sigma \, EF \, . \, DE = 0,$$

A C B will be the required equilibrium polygon. From the reasoning of § 37, it will be proper to divide the areas above the springing line A B by *equidistant* vertical lines, moderately near together, scale off the quantities corresponding to E F and D E, and find the proper position of A C B by one or two trials. It can thus be located with all desirable accuracy, as a slight movement of the point C vertically alters the quantities to be computed very materially. The reader who is not familiar with the higher mathematics can thus verify the results we are about to obtain.

Since C G may be considered the unknown quantity by which to locate A C and B C, its value may be deduced from the above equation. Let the half-span A K, $=$ K B, $= c$; the height or rise of the arch at the crown $= k$; the distance K G, from mid-span to the position of the single weight, $= b$;

and the required maximum ordinate C G $= y_0$. Then will the value of C G be

$$y_0 = \frac{32}{5} k \frac{c^2}{5 c^2 - b^2},$$

which becomes, if $b = n c$, where $n =$ a fraction of the half-span,

$$y_0 = \frac{32}{5(5 - n^2)} k, \quad (1.)$$

a quantity independent of the span of the arch.

39. **Proof of Formula.** — Let A D, the distance from the abutment A to any ordinate D E, between A and G, $= x$. A G $= c + b$; G B $= c - b$. Since the ordinates to a parabola from the line A B are proportional to the product of the segments into which they divide the span, we have

$$\text{D E} : k = x(2c - x) : c^2, \text{ or D E} = \frac{k}{c^2}(2cx - x^2).$$

Also,

$$\text{D F} : y_0 = x : c + b, \qquad \text{or D F} = \frac{y_0}{c + b} x.$$

The required condition is that

$$\Sigma \text{ E F . D E} = 0, \text{ or } \Sigma (\text{D E} - \text{D F}) \text{ D E} = 0;$$

therefore,

$$\Sigma \text{ D E}^2 = \Sigma \text{ D F. D E.} \quad (1.)$$

(From the above expressions we see, that, if the area included between the rib and A B is considered positive, the area of the triangle A C B, superimposed upon it, will be deemed negative as before explained in Fig. 14.)

Substituting the values of the lines from above in (1.), multiplying by dx, and writing the sign of integration, we get for the left-hand member,

$$\int_0^{2c} \frac{k^2}{c^4}(2cx - x^2)^2 dx = \frac{k^2}{c^4} \cdot \int_0^{2c} (4c^2x^2 - 4cx^3 + x^4) dx$$

$$= \frac{k^2}{c^4}\left(\tfrac{4}{3}c^2x^3 - cx^4 + \tfrac{1}{5}x^5\right)_0^{2c} = \tfrac{16}{15} k^2 c. \quad (2.)$$

For the right-hand member, between A and G, we get

$$\int_0^{c+b} \frac{y_0}{c+b} x \cdot \frac{k}{c^2}(2cx - x^2) dx = \frac{k y_0}{c^2(c+b)} \cdot \int_0^{c+b} (2cx^2 - x^3) dx$$

$$= \frac{k y_0}{c^2(c+b)}\left(\tfrac{2}{3}cx^3 - \tfrac{1}{4}x^4\right)_0^{c+b} = \frac{k y_0}{c^2}[\tfrac{2}{3}c(c+b)^2 - \tfrac{1}{4}(c+b)^3]. \quad (3.)$$

For the portion between G and B, if we write $c - b$ for $c + b$, and reckon x from B to the left, D F will equal $\frac{y_0}{c - b} x$, while D E will be unchanged;

so that the integration for the right-hand member of (1.), between G and B, and between the limits $x = 0$ and $x = c - b$, will give, simply by writing $-b$ for $+b$,

$$\frac{k y_0}{c^2}\left[\tfrac{2}{3} c\,(c-b)^2 - \tfrac{1}{4}(c-b)^3\right]. \quad (4.)$$

These two portions (3.) and (4.), for the right-hand member of (1.), being added together, will produce, when the terms with the odd powers of b are cancelled,

$$\frac{k y_0}{c^2}\left(\tfrac{5}{6} c^3 - \tfrac{1}{6} c\, b^2\right).$$

Finally equate this value with (2.) to satisfy (1.), and

$$\frac{k y_0}{6 c}(5 c^2 - b^2) = \tfrac{16}{15} k^2 c; \text{ or } y_0 = \tfrac{32}{5} k \frac{c^2}{5 c^2 - b^2}, \quad (5.)$$

which is the desired value of C G in terms of the constant quantities, and the variable distance K G. This expression is plainly applicable to points on either side of K.

40. **Formula for Horizontal Thrust.** — For any position of the weight, plot the value of y_0, and draw the equilibrium polygon. Then draw two lines from the extremities of the load line W, parallel to the sides of the polygon, and thus determine H, and the two vertical components of the reactions, which vertical components will be the same as for a beam supported at its ends. But, from the simple relations of the similar triangles A G C and 0 3 1, Fig. 8, as also B G C and 0 3 2, we may write a general formula for H, if desired. Thus we have

$$y_0 : c - b = P_2 : H, \qquad \text{or } P_2 = \frac{y_0}{c - b} H;$$

$$y_0 : c + b = W - P_2 : H, \text{ or } W - P_2 = \frac{y_0}{c + b} H.$$

Eliminating P_2 in the second equation, by substituting its value from the first one, we get

$$W - \frac{y_0}{c - b} H = \frac{y_0}{c + b} H, \text{ or } (c^2 - b^2)\,W = 2 c\, y_0 H;$$

$$H = \frac{c^2 - b^2}{2 c\, y_0} W = \frac{1 - n^2}{2} \cdot \frac{5(5 - n^2)}{32} \cdot \frac{c}{k} W.$$

This value also will apply to a load on either side of the centre.

It will be observed that, to obtain this value of H, we have simply to divide $\frac{1}{2}(1-n^2)$ by the factor which multiplies k in (1.), § 38, to obtain the variable factor here.

41. **Computation of y_0 and H.** — The numerical values of these factors are worth obtaining, as, the computations once made, the results apply to every parabolic rib with pivoted ends. Let the span of the arch be divided into any convenient number of equal parts, and, for illustration, suppose that the number is ten, as shown in the figure; let a weight W be placed successively over each point of division, being supported by the rib. The calculation may conveniently proceed in the following manner: —

Find the different values of y_0 for different positions of W, by equation (1.), § 38. Then compute H by § 40. The calculation and results are given below; the equilibrium polygons and values of H for one-half of the arch are represented in Fig. 8. As n^2 is positive, whether n is + or —, the values of y_0 and H will be symmetrical on each side of the centre.

VALUES OF y_0 AND H.

$n=\frac{b}{c}$	=	0	0.2	0.4	0.6	0.8
$5-n^2$	=	5.00	4.96	4.84	4.64	4.36
$5(5-n^2)$	=	25.00	24.80	24.20	23.20	21.80
$\frac{32}{5(5-n^2)}$	=	1.280	1 2903	1.3223	1 3793	1.4679.

Multiply these factors by k to give y_0.

$\frac{1}{2}(1-n^2)$	=	0.50	0.48	0.42	0.32	0.18
$\frac{1}{2}(1-n^2)\div\frac{32}{5(5-n^2)}$	=	0.3906	0.3720	0.3176	0.2320	0.1226.

Multiply these factors by $\frac{c}{k}$ W to give H.

For any other desired division of the span, proceed in a similar way.

42. **Remarks.**—If every point of division were loaded with W at the same time, the value of the horizontal thrust would be equal to the sum of the H's for each load, that is, the factor in column 0 plus twice each of the others, and the sum multiplied by the factor $\frac{c}{k}$ W; we thus obtain $2.479 \frac{c}{k} W = H$. If a *truss* were uniformly loaded horizontally, the bending moment at the middle would be one-eighth of the total load multiplied by the span, or, for a truss of ten panels, with W = one panel load,

$$M = \frac{10\,W \,.\, 2c}{8} = 2\tfrac{1}{2}\,c\,W;$$

and the tension in the lower chord, or the compression in the upper chord, would be found by dividing this quantity by the height of the truss, k. If the span of the arch just treated had been divided into twenty equal parts, the value of H, for loads at all the points of division, would have been $4.990 \frac{c}{k} W$. The truss, as before, would give $\frac{20\,W \,.\, 2c}{8\,k} = 5 \frac{c}{k} W$.

We thus see that the equilibrium polygon, for a number of equal loads, equidistant horizontally, on a parabolic rib, gives a value of H approximating closely to that for a uniform load on a truss of height k, coming nearer as the loads increase in number, and agreeing when the load is continuous. Then the equilibrium polygon becomes a curve, coinciding perfectly with the parabolic rib, and gives the horizontal thrust to which we are accustomed in the bowstring girder under a maximum load.

43. **Computation of Bending Moments.**—While the ordinates can be readily scaled from a diagram, one who wishes may compute values of the bending moment M for numerous points, when W is placed on any one point. If y denotes the ordinate from A B to the inclined line, and z the ordinate of the parabola from any point D, the bending moment may be written,—

$$M = H\,(y - z).$$

If put in this form, it will be seen, that, in the neighborhood of y_0, M will be positive, coinciding with the moments for a beam supported at its two ends. As this is the most familiar flexure of a beam or truss, we have chosen to consider it as positive: § 12. The ordinates y and z can be readily calculated from the figure. Thus, if the weight is at 0.4 c from the middle of the span, we have found y_0 to be 1.3223 k. If the span is divided into ten parts, the number of divisions on one side of the weight being seven, y will be successively $\frac{1}{7}$, $\frac{2}{7}$, $\frac{3}{7}$, &c., of y_0; on the other side y will be $\frac{1}{3}$ and $\frac{2}{3}$ of y_0. The sum of the denominators always equals the number of divisions, and the fractions increase from both ends up to unity. After finding the first y at each end, we get the others by simple addition, and the row is checked by obtaining y_0 at the proper point. As stated in § 39, the ordinate z is proportional to the product of the segments into which it divides the span; or, if it is at a distance $n\,c$ from the middle, we have,

$$z = (1 + n)\,c\,(1 - n)\,c\,\frac{k}{c^2} = (1 - n^2)\,k.$$

The factors by which k is to be multiplied can therefore be at once obtained by taking the *decimals* which are found in the second line of the table for y_0, § 41.

The computations may then be set down in the following shape, viz.: —

VALUES OF M.

Point of Division.	1	2	3	4	5	6	y_0 7	8	9	
$\frac{n}{7}y_0$ =	.1889	.3778	.5667	.7556	.9445	1.1334	1.3223	.8815	.4408	$= \frac{n}{3}y_0$.
z =	.36	.64	.84	.96	1.00	.96	.84	.64	.36	k
$y - z$ =	−.1711	−.2622	−.2733	−.2044	−.0555	+.1734	+.4823	+.2415	+.0808	k
Multiply by $H = 0.3176 \frac{c}{k}$ W.										
M =	−.0543	−.0833	−.0868	−.0649	−.0176	+.0551	+.1532	+.0767	+.0257	c W

With the explanation already given, this table will be understood. The letter y_0 is placed over 7 as a convenience, to show that the value y_0 occurs at this point of division. If the load is on the right of the centre, these numbers run from the left abutment; if the load is on the left of the centre, they must be reckoned from the right abutment.

44. **Table of Bending Moments.** — We have carried out this computation for a load at each joint successively, the span being divided into ten equal parts, and have prepared a table given on p. 53. A table for a span divided into twenty parts may be found in "Engineering News," Vol. IV. p. 108. As a load on either side of the middle gives the same set of values in the reverse order, it is necessary to calculate but one-half of the table.

As many decimals may be taken as will give sufficiently accurate results. By the use of logarithms the labor of preparing another table for a different number of divisions is very little. Each column belongs to the point of division whose number stands at its top, the numbers commencing at the left abutment. Each horizontal line contains the factor for bending moment at each point of division for a load W on the point marked at the beginning of the line. The values of H are placed for convenience in the last column.

It is worthy of notice, that, while the value of y_0 is independent of the span of the arch, M is independent of the height of the arch. As it was proved, in § 28, that the parabola is the equilibrium curve for a load distributed uniformly horizontally, this arch ought to be very nearly in equilibrium when we place at once on each one of the nine points a load W: by footing up the vertical columns of the table we shall find but a very small residual moment at each joint.

45. **Interpolation.** — In the solution of a particular example, it may happen that the points at which the weight will be concentrated will not coincide with the points of division which we have taken. It will then be necessary to determine new values of y_0 and H, which may be done by the original formulæ or by interpolation. A new table of M may then be calculated, values may be interpolated in the one given here, or, if preferred, from the value of H, and the vertical components of the reactions, we may draw an equilibrium curve for any combination of loads. The table here given, if not directly applicable in all cases, serves two purposes; one to show how a similar table can be made, and the other to indicate, by inspection, what arrangement of loads on any arch will produce the maximum bending moments.

If the successive values of any quantity increase at a tolerably uniform rate, any intermediate value between two given ones may be found by simple proportion. Otherwise we may use the formula for interpolation, —

$$\text{Desired quantity} = a + f\,[D_1 - \tfrac{1}{2}\,(1 - f)\,D_2],$$

in which a denotes the first given quantity, f the fraction of a division from a to the desired quantity, and D_1 and D_2 the *first* and *second differences*. To illustrate, take the values of H in § 41. If we place these in a column as below, find the amount

b.	H.	D_1.	D_2.
0	.3906		
		—.0186	
.2*c*	.3720		—.0358
		—.0544	
.4*c*	.3176		—.0312
		—.0856	
.6*c*	.2320		—.0238
		—.1094	
.8*c*	.1226		

of *increase* from quantity to quantity, and then subtract these differences from one another, marking each + if it is an increment, and *vice versa*, we obtain the columns of first and second differences as marked. Now suppose that we wish to determine a value of H at $b = .5\,c$; a will be .3176, $f = \frac{1}{2}$, $D_1 = —.0856$, and D_2 for an average value between .0312 and .0238, $= -.0275$. If we substitute in the formula, it then becomes

$$H\ (\text{for } .5\,c) = .3176 + \tfrac{1}{2}\,[—.0856 — \tfrac{1}{2} \cdot \tfrac{1}{2}\,(—.0275)]$$
$$= .3176 + \tfrac{1}{2}\,(-\,.0856 + .0069) = .2783.$$

The factor for y_0, at one-third of the interval between .4 *c* and .6 *c*, will, in the same way, be

$$1.3223 + \tfrac{1}{3}\,[.0570 - \tfrac{1}{2} \cdot \tfrac{2}{3}\,(.0283)] = 1.3382.$$

Careful heed must be paid to the signs.

46. **Examples.** — It will help to fix the ideas, if we draw an equilibrium polygon for some combination of weights. We shall take but a few loads, in order to have the diagram clear; but the reader may vary the example by taking other amounts in other places. The values of the two vertical components of the abutment reactions will be the sums of the components for each load, and the amount of H for the whole load will be the sum of the separate H's. Multiply each numerical factor which belongs to H by the number of units of weight which are

placed on the point to which the factor refers, add up the products, and plot the resulting value of H horizontally from the point of division on the load line between the two vertical components of the reactions.

For example: Let us draw the equilibrium polygon for an arch of 100 feet span, 20 feet rise, whose weight is at present, for simplicity's sake, neglected, when it is loaded with weights of 3 tons, 2 tons, 4 tons, and 2 tons, at the end of the 3d, 6th, 8th, and 9th division from the left, of ten equal horizontal divisions, as shown in Fig. 9, where the numbers denote the weights and the points of division above mentioned. The supporting force on the left will be

$$P_1 = \frac{2 \times 1 + 4 \times 2 + 2 \times 4 + 3 \times 7}{10} = 3.9 \text{ tons.}$$

$$\therefore \quad P_2 = 7.1 \text{ tons.}$$

From the table for H,

$$H = (0\ 3176 \times 3 + 0.372 \times 2 + 0.232 \times 4 + 0.1226 \times 2)\, \tfrac{50}{20}$$
$$= 2.87 \times \tfrac{5}{2} = 7.175 \text{ tons.}$$

These quantities are plotted in the stress diagram, as seen in the figure, and the equilibrium polygon is then drawn. The reader who reproduces this figure, or draws another, can be assured of the accuracy of the construction by the closing of the equilibrium polygon on the point of support. The weight of the arch itself may be accounted for by concentrating the proper amount at each point of division. Such amounts will increase towards the springing in proportion to the square of the secant of inclination to the horizon; for we recall the fact that the parabolic rib is to increase in breadth from crown to springing, and the amount in length projected into a horizontal foot increases in the same way. The weight of each division of the arch can be obtained with sufficient accuracy from a moderately large figure.

Another good construction is the curve for a uniform load over one-half of the span. The equilibrium curve for such a load, on the left half of Fig. 8, is represented in that figure; the work may be carried out in detail by the reader, and compared with the same curve for the three-hinged rib.

47. **Numerical Value of M.** — It will be seen that the polygon and rib of Fig. 9 approach quite nearly at 3. We can find the distance between them vertically, if we wish, from the table of M. The bending moment will be, taking the column 3,

$$M = 50\,(+.153 \times 3 - .073 \times 2 - .075 \times 4 - .043 \times 2) = -3.650 \text{ ft.tons.}$$

$$\frac{M}{H} = \frac{-3\ 65}{7.2} = -0.5 \text{ ft.} = y - z.$$

A similar operation may be performed at any other point.

48. **Shear Diagram.** — This investigation of shear is intended to apply to ribs of an I-section or to those framed with open-work or skeleton webs, and not to those of solid section, rectangular, circular, or otherwise, nor to stone arches: in these latter classes the shearing forces need seldom be taken into account.

Adhering still to the case of a single weight W, at a distance b from the middle of the span, we found that the vertical component, P_2, of the reaction at the end nearest to the weight, would be $\frac{c+b}{2c}$ W, and at the other end $\frac{c-b}{2c}$ W. As seen in Fig. 8, the diagram for shear on a beam will be, if we take the shear on the *left* of any section, $a\,d = P_1, = 3\text{–}1$, on the left of the weight, and $l\,g = -P_2, = 3\text{–}2$, on the right of the weight, giving the two rectangles included between $a\,l$ and the broken line $d\,e\,f\,g$. As the parabola is in equilibrium under a load of uniform intensity horizontally (§ 28), in which case there will be no bracing required, — no shear for any bracing to resist, — it is manifest that the diagram for that portion of the shear which is here carried, at each *vertical* section, by the flanges or chords, must be similar to the shear diagram for a uniform load on a beam supported at both ends; that is, to such a figure as $a\,i\,m\,n\,l$. If, then, we can determine the value of $a\,i$, or of the equal ordinate $l\,n$, we can draw this portion of the figure.

It is a well-known property of the parabola, that a tangent at the springing of the arch will intersect the middle ordinate at a distance k above the crown, equal to the rise of the arch. If, then, we draw a line 0–4 in the stress diagram, parallel to the tangent A L, drawn as just described, the distance 3–4, intercepted on the vertical line, will be the amount of vertical force necessarily combined with H to give a thrust coinciding with the rib at the springing point. Lay off, therefore, 3–4 at *a i*, and an equal amount at *l n*; then draw the straight line *i n*, cutting *a l* at its middle point *m*: the ordinates to this line from *a l*,

Parabolic Rib, Hinged at Ends.

§ 44. $M = m c W$. Values of m.

	1	2	3	4	5	6	7	8	9	H
W on 9	—.024	—.039	—.043	—.038	—.023	+.002	+.037	+.082	+.136	$.123 \frac{c}{k} W.$
" 8	—.044	—.068	—.075	—.063	—.032	+.017	+.085	+.171	+.076	.232 "
" 7	—.054	—.083	—.087	—.065	—.018	+.055	+.153	+.076	+.025	.318 "
" 6	—.054	—.078	—.073	—.037	+.028	+.123	+.047	+.002	—.014	.372 "
" 5	—.041	—.050	—.028	+.025	+.109	+.025	—.028	—.050	—.041	.391 "
" 4	—.014	+.002	+.047	+.123	+.028	—.037	—.073	—.078	—.054	.372 "
" 3	+.025	+.076	+.153	+.055	—.018	—.065	—.087	—.083	—.054	.318 "
" 2	+.076	+.171	+.085	+.017	—.032	—.063	—.075	—.068	—.044	.232 "
" 1	+.136	+.082	+.037	+.002	—.023	—.038	—.043	—.039	—.024	.123 "

§ 53. $V = n W$. Values of n.

	1	2	3	4	5	6	7	8	9	
W on 9	—.121	—.072	—.023	+.026	+.075	+.125	+.173	+.223	+.272	—.678
" 8	—.218	—.125	—.032	+.061	+.153	+.247	+.339	+.432	—.475	—.382
" 7	—.272	—.145	—.018	+.109	+.236	+.364	+.491	—.382	—.255	—.128
" 6	—.270	—.121	+.028	+.177	+.325	+.475	—.377	—.228	—.079	+.069
" 5	—.204	—.047	+.109	+.265	+.422	—.422	—.265	—.109	+.047	+.204
" 4	—.069	+.079	+.228	+.377	—.475	—.325	—.177	—.028	+.121	+.270
" 3	+.128	+.255	+.382	—.491	—.364	—.236	—.109	+.018	+.145	+.272
" 2	+.382	+.475	—.432	—.339	—.247	—.153	—.061	+.032	+.125	+.218
" 1	+.678	—.272	—.223	—.173	—.125	—.075	—.026	+.023	+.072	+.121

at all points, will represent the amount of vertical force to be combined with the horizontal thrust to put the rib in equilibrium. The remaining ordinates are drawn at the middle of each division; and, where the amount subtracted is greater than the original shear, the remainder will be of the opposite sign. The signs are placed in the areas of this figure; and it will be apparent that the ordinates are reckoned from the inclined line *i n*, all *above that line* in our figure representing *positive* or upward shear on the *left* of a *vertical* plane of section, while those *below i n* will be *negative.* See p. 31.

49. **Shear on a Normal Section.** — To obtain the shear on a right or normal section, as at Q, we must draw a line *q s* parallel to the normal section at Q, and project *r q* upon it, thus finding *s q* as the shear at Q. A similar construction will determine the shear at any other point. The property of the [illegible] above alluded to makes it easy to find the direction [illegible] which will be perpendicular to a tangent at Q; a tangent at Q will strike K L at S, a distance above the crown equal to that of the extremity R of the horizontal line Q R below it. What has been done by the above steps may also be easily seen from the sketch above Fig. 8. At A, P_1 will be *a d* or 3–1, and the whole vertical force to be combined with H will be *a i* or 3–4, which when subtracted from *a d* leaves *i d* or 4–1 as the negative shear on a vertical plane, and F, *t d*, or 6–1, as the shear on a right section at A.

In treating any arched rib, we shall desire to find the maximum shear at any section produced by a combination of weights at several points. It will be easier to find the sum of the several shears on a vertical section from single weights, and then find the normal component once for all, than to resolve each vertical shear separately; hence the shear diagram of Fig. 8 and of subsequent figures will simply show the shears on the several vertical sections before they are projected on the normal sections.

50. **Formula for Vertical Shear.** — A formula for this vertical shear may be deduced without difficulty. If Y is the ordinate to *i n* from any point of *a l*, and Y_1 its value at the springing, we have from the statement of the last section,

$$Y_1 : H = 2k : c, \text{ or } Y_1 = \frac{2k}{c} H.$$

The vertical shear V in the web, at the abutment on the left, will then be,

$$V = P_1 - Y_1 = \frac{c \pm b}{2c} W - \frac{2k}{c} H. \quad (1.)$$

For successive points, P_1 will remain the value of the original shear until we pass the weight, when it will become $P_1 - W$ or $-P_2$. Y will diminish at a constant rate; and, if we deduct at each point the ordinate from *a l* to the inclined line, we shall get the desired results.

51. **Computation of Shear.** — As an example we will find the vertical shear *midway* between the points of division of the arch of Fig. 8 with the load there shown.

$$P_1 = 0.3\,W; \; P_2 = 0\,7\,W; \; H = .3176 \frac{c}{k} W; \; Y_1 = .6352\,W.$$

This value of Y_1 is applicable to any parabolic arch with hinged ends, since it involves neither c nor k. Y at the middle of the first space $= \left(.635 - \frac{.635}{10}\right) W = .572\,W$; for every succeeding ordinate it diminishes $\frac{.635}{5} W$.

VALUES OF V.

Space.	1	2	3	4	5	6	7	8	9	10	
P_1	.3	.3	.3	.3	.3	.3	.3	−.7	−.7	−.7	$-P_2$
Y	.572	.445	.318	.191	+.064	−.064	−.191	−.318	−.445	−.572	
P − Y	−.272	−.145	−.018	+.109	+.236	+.364	+.491	−.382	−.255	−.128	W.

Three decimal places here will be as exact as four in the values of M. It will be seen by the ordinates in the shear diagram of Fig. 8, how the signs change.

52. **Remarks on Shear.** — We repeat that, as P_1 was taken as positive, the signs of the shears apply to the left side of each vertical or each normal section. In Fig. 10 the sketch marked R is an instance of positive shear, which acts up or outward on the left of the imaginary section and inward on the right of the same section. From the way in which the two parts of the arch will tend to slide at the section, we see that at R a tie will be required sloping down from the upper chord to the right (or a strut in the opposite direction), while negative shear, as represented in the sketch marked S, calls for a tie in the reverse direction.

53. **Table of Shears.**—A table has been computed by the preceding process, for shears at the middle points of ten equal spaces, into which the span is divided. It is intended to supplement the previous table of bending moments, and will serve as a guide for the calculation of any table with a greater or less number of spaces. It will be found on p. 53. A shear at a joint can be found, if desired, by taking the mean of two adjacent shears just obtained. It is easy to select from this table that combination of loads which will give on any parabolic arch, hinged at the ends only, the maximum shear of either kind in any one division, one arrangement being the complement of the other. These shears, as should be the case, foot up very nearly to zero for an equal load on every joint. It is only necessary to calculate one-half of the table; the other half will contain the same numbers in the reverse order, with the opposite signs. A table for an arch of twenty divisions was printed in "Engineering News," vol. iv., p. 124.

54. **Extent of Load to Produce Maximum Bending Moments and Shears.**—In single-span trusses the maximum bending moments, and consequently the maximum stresses in the chords, occur when the bridge is entirely covered with the live load; and the greatest shear at any section, or the greatest stress in any brace, exists when the bridge is covered with live load over one or the other, usually the longer, of the two segments into which the section divides the span. A simple inspection of the tables for M and V, lately given, will show that such rules are not true for an arch. Why this is so, will be seen, if we consider the fact that the portion of the arch, Fig. 8, between B and the point where C A crosses the rib, is under a bending moment of the positive kind, when there is a single weight at I, while from that point to A bending moments of the negative kind exist; and that an addition of another load near I will increase in amount most of the positive and negative moments, while one placed on the left half of the arch will have an opposite effect. The shearing forces for the braces, depending upon the change of stress in the flanges, will also be affected in the same way.

While an inspection of Fig. 8 will show, as was pointed out with regard to Fig. 4, in § 32, the extent of load to produce the maximum bending moment at any one point, and while the load to produce maximum shear at the same point can also be ascertained by inspection, § 15, an attempt has been made to represent, by the horizontal lines in the diagram, Fig. 11, those positions of the live load, or the extent of the loaded portion, which will give the maximum moments of both kinds at each of nineteen points of division represented in the figure, and also that arrangement of the live load which gives the maximum shear of either kind at the middle of each division. The full line denotes the loaded portion of the span when the maximum positive moment occurs at that point whose number is placed at the end of the line, positive being understood to mean that kind of moment which would make a previously straight beam concave on the upper side; and the remaining portion of the span must alone be covered with the live load to produce the maximum negative moment at the same pc Thus the maximum positive bending moment at 2, and 3 also, is found when the load is on all points from the left to 7 inclusive. A load from 8 to the right abutment gives the maximum —M. The maximum +M at 11 occurs when the arch is loaded from 9 to 14 inclusive.

The extent of live load required to produce the greatest upward, or positive, shear on the left of a section through the web or brace in any division, is indicated by the broken line drawn in its proper space; and a load over the complementary blank portion will give the maximum shear of the opposite kind in the same division. Thus the maximum +F, at the middle of 3–4, is found when the load extends from 4 to 9 inclusive; and the maximum —F, at the same place, when the load reaches from 1 to 3 and 10 to 19 inclusive. As a partial load, not extending to either abutment, will give the greatest M at some points, and as the same thing is true of the values of F, those writers who determine the greatest stresses by the usual test for maximum applied to an algebraic equation, which contains the expression for load as continuous from one abutment, must err in their results.

55. **Resultant Maximum Stresses.** — The steady or fixed

load, unless distributed uniformly horizontally, gives some definite bending moment and shear, of one sign or the other, at each point; and these amounts must first be obtained from the tables or by diagram. If, at a given point, the bending moment from fixed weight is +, the arrangement of rolling load which gives the maximum +M at that point will conspire with the steady load, and give an actual maximum +M; while that arrangement of rolling load which, in itself, gives a maximum −M, will reduce the moment from steady load. If large enough to prevail against the +M, the rolling load will produce an actual maximum −M; but, if not, it will only cause a minimum +M. Similar remarks might be made concerning shear.

An absolute maximum M of either kind, for a uniform load, will be found, if we sum up the quantities in the table, to occur at the middle of the half-span. The loads to produce these values are seen in Fig. 11. The absolute maximum ±F is found at the abutments, while another value, nearly equal in amount, occurs at the crown. These absolute maxima are found by comparing footings of the several columns, p. 53.

If Fig. 10 is supposed to represent a portion of the rib of Fig. 8 or Fig. 12, the web system being of any type or a continuous plate, we shall find that, when the chords or flanges lie on the opposite sides of any equilibrium polygon, they will be in compression from the weight which belongs to that polygon. When they both lie on the same side, the nearer chord or flange will be in compression and the farther one in tension. Hence the extent and amount of load to produce maximum stress of either kind in any chord piece can be found by inspection.

The actual stress is found by taking moments about the proper joint in the opposite chord, as is done in bridge trusses, using either H multiplied by the vertical ordinate, or the thrust in the side of the equilibrium polygon multiplied by the length of the perpendicular, drawn from the joint to that side, as may be preferred, and dividing by the length of the perpendicular from the same joint to the chord piece in question, considered as straight between its two joints. In this way the stress resulting from the direct thrust combined with the bending moment is at once determined.

Again, imagine a right section made in Fig. 8, through any panel like Fig. 10, and arrow-heads placed on the equilibrium polygons on the left of, and thrusting against the section. If the forces represented by such arrows have components acting up or outward along the section, they will cause positive shear in the web at that section; if such components act inward, they will cause negative shear. Hence the extent of load to produce maximum shear of either sign in a particular panel can also be found by inspection, and the amount of that shear can then be determined.

56. **Example of Flange Stresses.**—It may be instructive to make a little numerical calculation for the rib of Fig. 9, 100 feet span and 20 feet rise, supposing it to be loaded with the four weights only which are shown in the figure. The maximum positive moment is plainly at 8. If the rib is made of a web and two flanges $2\frac{1}{2}$ feet from centre to centre, what will be, with this load, the stress in each flange at 8? If our figure were larger, we could scale the ordinate above 8, and get the bending moment directly; but, as the sketch is small, we will refer to the table. We thus find that

$$M = (.082 \times 2 + .171 \times 4 + .002 \times 2 - .083 \times 3)\ 50 = 30.15 \text{ foot tons.}$$

From the same table we find that

$$H = (.123 \times 2 + .232 \times 4 + .372 \times 2 + .318 \times 3)\ \tfrac{50}{20} = 7.18 \text{ tons.}$$

Then $30.15 \div 7.18 = 4.2$ feet, ordinate at 8. If we call the vertical depth of the rib at 8, three feet, the whole ordinate to the lower flange will be $4.2 + 1.5 = 5.7$ feet, and to the upper flange $4.2 - 1.5 = 2.7$ feet. The compression in the upper flange will be $7.18 \times 5.7 \div 2.5 = 16.37$ tons; and the tension in the lower flange $7.18 \times 2.7 \div 2.5 = 7.75$ tons.

Draw 0–5 parallel to the tangent at 8. Drop perpendiculars 3–6 and 4–7 on it from 3 and 4. On a right section close to, but on the left of 8, there will be positive shear 4–7, equal to 2.1 tons. On the right of 8 will be found 3–6, or 1.5 tons negative shear, to be resisted by the web.

CHAPTER IV.

PARABOLIC RIB WITH FIXED ENDS.

57. **Values of Ordinates.**—Passing next to the parabolic arch, fixed at the ends, we recall, from § 16, that, to locate the equilibrium polygon for a single load at any point, we need *three ordinates*, one at each end, and the third passing through the weight, and that the three conditions by which these must be obtained are, 1st, that the change of span is zero; 2d, that the change of inclination at the abutments is zero; and, 3d, that the abutment deflection is zero. As expressed in the notation used, the three equations of condition are

$$\Sigma\, \mathrm{E\,F\,.\,D\,E} = 0,$$
$$\Sigma\, \mathrm{E\,F} = 0,$$
$$\Sigma\, \mathrm{E\,F\,.\,D\,B} = 0.$$

If, in Fig. 12, I N L represents the desired equilibrium polygon for a weight W, attached to the rib A Q B at a point distant T G, $= b$, horizontally from the middle of the span; and if the span A B $= 2\,c$, the rise of the arch $= k$, A I $= y_1$, G N $= y_0$, and B L $= y_2$, we will prove that

$$y_0 = \tfrac{6}{5}\,k, \quad (1.)$$

$$y_1 = \tfrac{2}{15} \cdot \frac{c + 5\,b}{c + b}\,k = \tfrac{2}{15}\,\frac{1 + 5\,n}{1 + n}\,k, \quad (2.)$$

$$y_2 = \tfrac{2}{15} \cdot \frac{c - 5\,b}{c - b}\,k = \tfrac{2}{15}\,\frac{1 - 5\,n}{1 - n}\,k, \quad (3.)$$

when $b = n\,c$.

58. Value of First Equation. — As before, the first condition may be written,

$$\Sigma\, EF \, . \, DE = \Sigma\, (DE - DF)\, DE = 0, \text{ or } \Sigma\, DE^2 = \Sigma\, DF \, . \, DE. \quad (1.)$$

If $AD = x$, $DE = \frac{k}{c^2}(2c - x)\, x$, as in § 39. $AG = c + b$; $GB = c - b$. If y_1 or y_2 becomes negative, it is to be laid off below AB, but otherwise above: the figure represents y_2 as negative; and, in the majority of cases, y_1 and y_2 have opposite signs. If a line be drawn horizontally from I, DF, as long as it is on the left of y_0, will be divided into a constant part y_1, and a remainder which varies with the distance from I. Hence we see that

$$DF = y_1 + \frac{y_0 - y_1}{c + b} x.$$

For the right-hand member of (1.), between A and G, we therefore get

$$\int_0^{c+b} \left(y_1 + \frac{y_0 - y_1}{c + b}\, x \right) \frac{k}{c^2} (2cx - x^2)\, dx =$$

$$\frac{k}{c^2} y_1 \int_0^{c+b} (2cx - x^2)\, dx + \frac{k}{c^2} \cdot \frac{y_0 - y_1}{c + b} \int_0^{c+b} (2cx^2 - x^3)\, dx =$$

$$\frac{k}{c^2} y_1 [c(c+b)^2 - \tfrac{1}{3}(c+b)^3] + \frac{k}{c^2}(y_0 - y_1)[\tfrac{2}{3} c (c+b)^2 - \tfrac{1}{4}(c+b)^3]. \quad (2.)$$

For the portion between G and B, if we write $c - b$ for $c + b$, and reckon x from B to the left, we get

$$DF = y_2 + \frac{y_0 - y_2}{c - b} x,$$

the sign of y_2 being contained in the symbol. Then the integration for the right-hand member of (1.), between B and G, or between the limits 0 and $c - b$, will give, when we substitute y_2 for y_1, and $c - b$ for $c + b$,

$$\frac{k}{c^2} y_2 [c(c-b)^2 - \tfrac{1}{3}(c-b)^3] + \frac{k}{c^2}(y_0 - y_2)[\tfrac{2}{3} c (c-b)^2 - \tfrac{1}{4}(c-b)^3]. \quad (3.)$$

The left-hand member of (1.) was shown to be, in § 39, (2.),

$$\int_0^{2c} \frac{k^2}{c^4} (2cx - x^2)^2\, dx = \tfrac{16}{15} k^2 c. \quad (4.)$$

The two portions, (2.) and (3.), of the right-hand member, being added together, when the coefficients of y_0, y_1, and y_2 are reduced, will be equated with (4.), the left-hand member of (1.), producing

$$\frac{k}{6c^2}\left\{y_0(5c^3-cb^2)+\tfrac{1}{2}y_1(c+b)^2(3c-b)\right.$$
$$\left.+\tfrac{1}{2}y_2(c-b)^2(3c+b)\right\}=\tfrac{16}{15}k^2c,$$

or

$$2c(5c^2-b^2)y_0+(c+b)^2(3c-b)y_1+(c-b)^2(3c+b)y_2=\tfrac{64}{5}kc^3. \quad (5.)$$

59. **Values of Second and Third Equations.** — It is not necessary to integrate in order to obtain equations from the other two conditions, although they may be derived quite simply in that way. The second condition may be written,

$$\Sigma\, \mathrm{E\,F} = \Sigma\,(\mathrm{D\,E}-\mathrm{D\,F}) = 0, \quad \text{or} \quad \Sigma\, \mathrm{D\,E} = \Sigma\, \mathrm{D\,F}.$$

The first member is the summation of all the ordinates to the arch, or the included area between the rib and the line A B. The area of a parabolic segment being equal to two-thirds of the rectangle of the same base and altitude, the area will be $\frac{2}{3} \cdot 2c \cdot k$, or $\frac{4}{3}ck$. The second member will be the summation of all the ordinates to the two inclined lines, or the area of the two trapezoids, giving

$$\tfrac{1}{2}(y_0+y_1)(c+b)+\tfrac{1}{2}(y_0+y_2)(c-b), \quad \text{or} \quad cy_0+\tfrac{1}{2}(c+b)y_1+\tfrac{1}{2}(c-b)y_2.$$

Equating the two values, we obtain the second equation,

$$2cy_0+(c+b)y_1+(c-b)y_2=\tfrac{8}{3}ck \quad (1.)$$

The condition that Σ E F . D B = 0, or that Σ (D E − D F) D B = 0, gives

$$\Sigma\, \mathrm{D\,E} \cdot \mathrm{D\,B} = \Sigma\, \mathrm{D\,F} \cdot \mathrm{D\,B},$$

and this condition is satisfied by the equivalent step of multiplying each area, just obtained, by the horizontal distance of its centre of gravity from one abutment, the right one for example, and equating the products. The left-hand member will then plainly be $\frac{4}{3}ck \cdot c$, or $\frac{4}{3}c^2k$. As the second expression above for the area of the trapezoids has three terms which correspond to the three triangles formed by drawing lines from N to A and B, we may multiply each triangle by the distance of its centre of gravity from B, obtaining

$$cy_0(c-\tfrac{1}{3}b)+\tfrac{1}{2}(c+b)y_1[c-b+\tfrac{2}{3}(c+b)]+\tfrac{1}{2}(c-b)y_2\tfrac{1}{3}(c-b),$$

or,

$$\tfrac{1}{3}cy_0(3c-b)+\tfrac{1}{6}(c+b)y_1(5c-b)+\tfrac{1}{6}y_2(c-b)^2.$$

Equating the two members, and clearing of fractions, we find that

$$2c(3c-b)y_0 + (c+b)(5c-b)y_1 + (c-b)^2 y_2 = 8c^3 k. \quad (2.)$$

60. **Solution of Equations.** — Equations (5.), § 58, and (1.) and (2.), § 59, contain the three unknown quantities. The eliminations may be performed as follows: —

Multiply (1.) by $c - b$, obtaining

$$2c(c-b)y_0 + (c+b)(c-b)y_1 + (c-b)^2 y_2 = (c^2 - bc)\tfrac{8}{3}k.$$

Subtract from (2.)

$$4c^2 y_0 + 4c(c+b)y_1 = (2c^2 + bc)\tfrac{8}{3}k. \quad (a.)$$

Multiply (2.) by $3c + b$,

$$2c(9c^2 - b^2)y_0 + (c+b)(15c^2 + 2cb - b^2)y_1 + (c-b)^2(3c+b)y_2 = (3c^3 + bc^2)8k.$$

Subtract (5.), and divide the remainder by $2c$,

$$4c^2 y_0 + 6c(c+b)y_1 = (\tfrac{7}{3}c^2 + bc)4k. \quad (b.)$$

Subtract $(a.)$,

$$2c(c+b)y_1 = (\tfrac{4}{15}c^2 + \tfrac{4}{3}bc)k, \quad \text{or } y_1 = \tfrac{2}{15} \cdot \frac{c+5b}{c+b}k.$$

Substituting this value in $(a.)$ or $(b.)$, we get

$$y_0 = \tfrac{8}{5}k,$$

and by analogy, or by substitution,

$$y_2 = \tfrac{2}{15} \cdot \frac{c-5b}{c-b}k.$$

61. **Remarks.** — The similarity between y_1 and y_2 is to be expected; for, when a load is moved from one side of the centre to an equal distance on the other, y_1 and y_2 change places. Therefore it must be remembered that y_2 is the value of the ordinate at that springing which is *nearer* to the weight. If

the load is in the middle, $b = 0$, and $y_1 = y_2$. It is worthy of notice that y_0 is a constant quantity for all positions of the weight. These ordinates can be easily computed for a weight at different points, and it will be seen that a value of b greater than $\frac{1}{5}c$ will make y_2 negative, or to be plotted below the springing line. The original reasoning showed, and the above equations will prove, that the third condition may be taken about the other abutment, and will still give the same values for the ordinates.

62. **Computation of Ordinates y_1 and y_2.**—If we propose to work out data for use with this type of arch also, we must first calculate the values of y_1 and y_2 for all points. Let a rib be divided into ten parts, equal horizontally as before; then, if $b = n\,c$, the results of the following table will be obtained. It

VALUES OF y_1 AND y_2.

$n = \frac{b}{c} =$	0	.2	.4	.6	.8	
$\frac{1+5n}{1+n} =$	$\frac{1}{1}$	$\frac{2.0}{1.2}$	$\frac{3.0}{1.4}$	$\frac{4.0}{1.6}$	$\frac{5.0}{1.8}$	
$\frac{2}{15} \cdot \frac{1+5n}{1+n} =$	0.1333	0.2222	0.2857	0.3333	0.3704	$k = y_1$.
$\frac{1-5n}{1-n} =$	$\frac{1}{1}$	$\frac{0}{0.8}$	$\frac{-1.0}{0.6}$	$\frac{-2.0}{0.4}$	$\frac{-3.0}{0.2}$	
$\frac{2}{15} \cdot \frac{1-5n}{1-n} =$	0.1333	0	−0.2222	−0.6667	−2.0	$k = y_2$.

is so similar to previous ones as to call for no explanation. Only remember that y_1 and y_2 change places for loads on the left of the crown. The equilibrium polygons for one half of the arch are shown in Fig. 12.

63. **Formulæ for H, P_1 and P_2.**—To obtain the value of H for a particular position of the load, we lay off y_1, y_0, and y_2 at A, G, and B, draw I N and N L, complete the stress diagram below, and draw 0–3 for H. The vertical components of the abutment reactions will be 2–3 and 3–1. If we draw the hori-

zontal dotted lines from I and L, we shall have similar triangles to those in the stress diagram, and may write

$$y_0 - y_1 : c + b = (2\text{–}3) : \mathrm{H}, \text{ or}$$

$$\mathrm{P}_1 = (2\text{–}3) = \mathrm{H}\frac{y_0 - y_1}{c + b} = \tfrac{8}{15} \cdot \frac{2 + n}{(1 + n)^2}\frac{k}{c} \cdot \mathrm{H},$$

$$y_0 + (-y_2) : c - b = (3\text{–}1) : \mathrm{H}, \text{ or}$$

$$\mathrm{P}_2 = \mathrm{W} - (2\text{–}3) = \mathrm{H}\frac{y_0 - y_2}{c - b} = \tfrac{8}{15} \cdot \frac{2 - n}{(1 - n)^2}\frac{k}{c} \cdot \mathrm{H}.$$

Substitute the value of (2–3) from the first equation, transpose, and obtain

$$\mathrm{H} = \frac{\mathrm{W}}{\dfrac{y_0 - y_1}{c + b} + \dfrac{y_0 - y_2}{c - b}} = \tfrac{15}{32} \cdot \frac{(c^2 - b^2)^2}{c^3 k} \cdot \mathrm{W} = \tfrac{15}{32}(1 - n^2)^2 \frac{c}{k}\mathrm{W}.$$

64. **Computation of Values.** — The amount of H for a load at any one point will then be found in the several columns of the table below. The first three values will be seen to be

VALUES OF H, P_1, AND P_2.

$n =$	0	.2	.4	.6	.8	
$1 - n^2 =$	1	.96	.84	.64	.36	
$(1 - n^2)^2 =$	1	.9216	.7056	.4096	.1296	
$\mathrm{H} =$	.4687	.4320	.3308	.1920	.0607	$\frac{c}{k}\mathrm{W}$.
$\mathrm{H}\frac{y_0 - y_1}{(1 + n)c} =$	0.5	0.352	0.216	0.104	0.028	W } = P.
$\mathrm{H}\frac{y_0 - y_2}{(1 - n)c} =$	0.5	0.648	0.784	0.896	0.972	W } = P.

greater, and the last two to be smaller, than the corresponding H's in § 41. It will next be necessary to find the vertical components of the reactions by multiplying H by the quantities noted in the last section: the results will be found in the last two lines. The larger value of P occurs at the nearer abutment. It will be noted that these quantities differ in amount from the two supporting forces of a single-span beam or truss.

If the H's for an equal load at each of the nine points of division are added together, we find that, for loads at all points,

$H = 2.4997 \frac{c}{k} W$, which agrees more closely with the amount for a truss or bowstring girder than did the value for a rib with hinged ends, § 42. It is due to the fact that the equilibrium polygon for a single weight crosses the rib oftener in the present case than in that of a rib with hinged ends; so that, when several loads are combined, the polygon will deviate from the parabola (the form of the rib, and the true equilibrium curve for a uniform distributed load) very little.

65. **Computation of Bending Moments.** — If, in place of scaling, we desire to compute the values of M in this case also, we may use the former equation, § 43,

$$M = H (y - z).$$

The values of the ordinates, z, to the parabola will be the same as before. If x denotes the distance from A to the foot of the ordinate y, and $x' =$ the distance from B to the foot of the same ordinate, in which case $x' = 2c - x$, we shall have

$$y = y_1 + \frac{y_0 - y_1}{c + b} x, \text{ on the left of the weight, and}$$

$$y = y_2 + \frac{y_0 - y_2}{c - b} x', \text{ on the right of the weight,}$$

the sign of y_2 being contained in the symbol.

Let us proceed to find the values of M, at both abutments and the nine other points, for a weight on the third point of division from the middle, towards the right. As above,

$$H = 0.192 \frac{c}{k} W; \quad \frac{y_0 - y_1}{c + b} = 0.5417 \frac{k}{c}; \quad \frac{y_0 - y_2}{c - b} = 4.6667 \frac{k}{c};$$

$$z = .36\,k,\ .64\,k,\ .84\,k,\ .96\,k,\ k,\ .96\,k, \text{ \&c., § 43.}$$

VALUES OF M.

									W			
$x =$	0 c	0.2 c	0.4 c	0.6 c	0.8	1.0	1.2	1.4	1.6	0.2 c	0 c	$= x'$
$\times .5417 \frac{k}{c}$	0	.1083	.2166	.3250	.4334	.5417	.6500	.7584	.8667	.9333	0	$\times 4.667 \frac{k}{c}$
$+ y_1$	.3333	.4416	.5500	.6583	.7667	.8750	.0833	1.0917	1.2000	.2667	—.6667	$+ y_2$
$z =$	0	.36	.64	.84	.96	1.00	.96	.84	.64	.36	0	k
$y - z =$	+.3333	+.0816	—.0900	—.1817	—.1933	—.1250	+.0233	+.2517	+.5600	—.0933	—.6667	k

Multiply by $H = 0.192 \frac{c}{k} W$.

M =	+.0640	+.0157	—.0173	—.0349	—.0371	—.0240	+.0045	+.0483	+.1075	—.0179	—.1280	c W

W is placed over the number of the point to which it is attached, and a double line is drawn on one side of W to denote the end of each series, running from the two ends of the table. The dividing line might just as well have been drawn on the left of W, if preferred. More frequent values of any of the preceding quantities may be obtained by interpolation, as explained before.

66. **Table of Bending Moments.** — A table of values of M has been prepared for this case of an arch with fixed ends, the span being divided into ten equal parts, and is here presented, p. 71. A table for twenty divisions may be found in "Engineering News," vol. iv., p. 178. At any one point, for a uniform load at all of the points of division, M reduces nearly to zero, as before. The greatest possible positive M, as well as the greatest possible negative M, for any combination of weights, occurs at each abutment; positive maximum when the span is loaded from the other abutment to and beyond the centre one point; negative when the other portion only of the span is covered. The load on the first point from the middle produces no M at the nearer abutment. There is another maximum at the third or seventh point, with loads nearly the reverse of the ones mentioned above. An inspection of the table will show these facts.

67. **Example.** — As soon as H, P, y_1, and y_2 have been obtained for all points, it is easy to draw an equilibrium polygon for any desired arrangement of load. Let us suppose that one must be constructed for weights of 2 tons, 6 tons, 3 tons, and 1 ton, on the 2d, 4th, 5th, and 8th points respectively, from the left abutment, of an arch of 100 feet span and 20 feet rise, Fig. 13, divided into ten equal parts along the span, as previously described. We will proceed as follows: —

The vertical components of the reactions cannot be computed for the load in the gross, as for a beam on two supports, but must be summed up from the values lately given. Referring to those data, we get

	P_1.	H.
2d joint,	$0.896 \times 2 = 1.792$ tons.	$0.192 \times 2 = 0.384 \frac{c}{k}$ tons.
4th "	$0.648 \times 6 = 3.888$ "	$0.432 \times 6 = 2.592$ "
5th "	$0.5 \times 3 = 1.500$ "	$0.469 \times 3 = 1.407$ "
8th "	$0.104 \times 1 = 0.104$ "	$0.192 \times 1 = 0.192$ "
	$P_1 = 7.284$ "	$H = 4.575$ "

$P_2 = 12 - 7.284 = 4.716$ tons. $H = 4.575 \times 2.5 = 11.44$ tons.

Since $H y_1 =$ moment at the springing A, Fig. 13; since each of these loads has a separate H and a definite y_1; and since the H's for the different loads all conspire to produce the total thrust, —we must calculate the arm with which the latter acts at one or both springings, that is, the ordinate y_1' or y_2' of the point whence the equilibrium polygon must start. We satisfy the equation

$$y_1' \cdot \Sigma H = \Sigma H \cdot y_1, \text{ or } y_1' = \frac{\Sigma H \cdot y_1}{\Sigma H},$$

which simply requires that the resultant moment shall be equal to the algebraic sum of the original moments. We therefore multiply each H for a given weight by its y_1, and divide the sum of the products by the total H. The calculation having been made, as here set down, we find that y_1' is equal to —.02 feet, a comparatively insignificant amount. It is well to compute y_2' also, as a check on the accuracy of the subsequent drawing, and it will be found to be +3.34 feet.

$$\begin{array}{rcrl}
y_1. & \text{H.} & \text{M.} & \\
-.667 \times & 0.384 = & -0.256\,c & \text{tons.} \\
0 \times & 2.592 = & 0 & \\
+.133 \times & 1.407 = & +0.188 & " \\
+.333 \times & 0.192 = & +0.064 & " \\
 & 4.575) & -0.004 & " \\
 & & -0.0009\,k. & \\
 & & 20 & \\
 & y_1' = & -0.018 & \text{feet.}
\end{array}$$

While we may seem to have carried out this example in too much detail, we are aware that inattention to apparently trivial points will sometimes cause trouble, and we have therefore given most of the work at full length. Now lay off the weights in order on the load line, plot P_1 and P_2, lay off H on the proper side, draw the usual radiating lines to the extremity of H, start below A, a distance $-y_1'$, and draw the equilibrium polygon with sides parallel to the inclined lines of the stress diagram, checking the polygon by the fact that it strikes the extremity of the calculated ordinate y_2'. Fig. 13 illustrates this example. The diagram for vertical shear is also shown below, and needs no explanation, as the construction is similar to previous cases. The dotted lines in the stress diagram determine the value of Y_1. It is quite noticeable in this figure, how the shear changes sign wherever the bending moment becomes a maximum.

68. **Table of Shear.** — To find the numerical value of the vertical shear, from which we may derive the normal components resisted by the braces of an arch with fixed ends, we proceed as we did in the case of an arch with hinged ends. The values of P_1, the vertical component of the abutment reaction at the left, have been found. We then need only calculate the value of $Y_1 = 2\frac{k}{c}H$, and form a table, as was done in § 51. It is not necessary to repeat the operations here. A table of shears for an arch with fixed ends, and for ten divisions, has been prepared, and is appended, p. 70. The same remarks apply to it as to the previous similar table for the parabolic arch with hinged ends. For a table for twenty divisions, see "Engineering News," vol. iv., p. 193.

69. **Extent of Load to produce Maximum M and F.** — A diagram is also presented, Fig. 15, showing, by the full lines, the loads required to produce the maximum +M, from live load, at the point whose number is attached to the line, and by the remaining blank portion the load required for maximum —M at the same point. The broken lines and the blank portion in each *space* represent the way of distributing the load for maximum +F and —F respectively. It is still more apparent from this figure than from Fig. 11, that any investigation which considers the rolling load as continuous from one

PARABOLIC RIB, FIXED AT ENDS.

§ 66. $M = mcW$. Values of m.

	0	1	2	3	4	5	6	7	8	9	10	H
W on 9	+.022	+.006	—.005	—.012	—.013	—.010	—.002	+.011	+.028	+.051	—.121	$.061 \frac{c}{k}$ W.
" 8	+.064	+.016	—.017	—.035	—.037	—.024	+.004	+.048	+.107	—.018	—.128	.192 "
" 7	+.095	+.019	—.031	—.054	—.050	—.020	+.036	+.119	+.028	—.036	—.073	.331 "
" 6	+.096	+.011	—.040	—.056	—.037	+.016	+.104	+.026	—.017	—.026	0	.432 "
" 5	+.062	—.006	—.037	—.031	+.012	+.094	+.012	—.031	—.037	—.006	+.062	.469 "
" 4	0	—.026	—.017	+.026	+.104	+.016	—.037	—.056	—.040	+.011	+.096	.432 "
" 3	—.073	—.036	+.028	+.119	+.036	—.020	—.050	—.054	—.031	+.019	+.095	.331 "
" 2	—.128	—.018	+.107	+.048	+.004	—.024	—.037	—.035	—.017	+.016	+.064	.192 "
" 1	—.121	+.051	+.028	+.011	—.002	—.010	—.013	—.012	—.005	+.006	+.022	.061 "

§ 68. $V = nW$. Values of n.

	0–1	1–2	2–3	3–4	4–5	5–6	6–7	7–8	8–9	9–10
W on 9	—.081	—.057	—.033	—.008	+.016	+.040	+.064	+.089	+.113	—.863
" 8	—.242	—.165	—.088	—.011	+.066	+.142	+.219	+.296	—.627	—.550
" 7	—.379	—.247	—.115	+.017	+.150	+.282	+.414	—.453	—.321	—.189
" 6	—.426	—.253	—.080	+.093	+.266	+.438	—.389	—.216	—.043	+.130
" 5	—.343	—.156	+.031	+.219	+.406	—.406	—.219	—.031	+.156	+.343
" 4	—.130	+.043	+.216	+.389	—.438	—.266	—.093	+.080	+.253	+.426
" 3	+.189	+.321	+.453	—.414	—.282	—.150	—.017	+.115	+.247	+.379
" 2	+.550	+.627	—.296	—.219	—.142	—.066	+.011	+.088	+.165	+.242
" 1	+.863	—.113	—.089	—.064	—.040	—.016	+.008	+.033	+.057	+.081

abutment over a portion of the span will not determine actual maximum stresses. See § 54.

70. **Comparison of Ribs; Fixed and Hinged at Abutments.** — A comparison of Fig. 15 with Fig. 11 will be instructive, as showing the different loading, when hinges are omitted, to produce maximum bending moments and shears. There are four points near the ends of the rib with fixed ends, which require that loads should be on both ends of the span at once, to produce the maximum +M at those points; and five points at the middle which have the maximum —M under similar circumstances. In some structures such conditions can be realized. If we foot up the plus and minus values of the columns in the tables for M and V, we shall readily see that, with the exception of the springing points, all the points in the arch with fixed ends have *less* maximum bending moment either kind, for a load W at each loaded point, than in the c of the arch with hinged ends, and, in most cases, the values a materially less. A similar comparison of maximum shears will show that the arch with fixed ends has to carry more shear over its web or bracing for all the divisions of the first and last quarters of the span, and less for the middle half of the span, than an arch with hinged ends. These considerations alone would indicate the superiority of the arch with fixed ends over the other type, as requiring less material in the flanges or chords, and throwing the heavier bracing towards the abutments; the value of the direct thrust, however, as indicated by the previously computed amounts of H, varies according to the amount of load, and conspires with the compression from bending moment, so that the sections of the two chords must be designed for the maximum compression and tension at all points; the effect of rise or fall of temperature will be shown to be greater on the rib with fixed ends, reqnring a greater increase of section to provide for it.

CHAPTER V.

CHANGE OF TEMPERATURE.

71. **Action of Change of Temperature.** — If the arch, when either fixed or hinged at the ends, is exposed to a change of temperature, it will tend to change its shape. If the rib were perfectly free, its expansion or contraction would be uniform in all directions, so that the new arch would be the old arch on a slightly altered scale. In a bowstring girder, the tie expands and contracts with the bow, so that the horizontal projection of the change of length of the bow is the same as the elongation or contraction of the horizontal member. But as the abutments of the arch are considered as fixed, its span must remain unchanged; and the alteration of the arch by a change of temperature will be manifested by a rise or fall of the crown of the arch, which movement, in the case of a metal rib, may be a marked quantity.

It is manifest, that, if we imagine the rib at its normal temperature to be placed upon its springing points or skewbacks, it will have a horizontal thrust against the abutments due to its form and weight. If the temperature changes, the structure endeavors to expand or contract in equal proportion in all directions; and hence, if possible, the span would be lengthened just in proportion to the rise of temperature t, the coefficient of expansion e, and the span $2c$, or the change of span would equal $2tec$. If t expresses the number of degrees of fall in temperature, it may be called *minus*, and the quantity $2tec$

will denote the shortening of the span. But this attempted change of length, being resisted at the points of attachment, cannot take place, but must cause a horizontal force, either tension or compression, which keeps the span invariable. This $+H$ or $-H$ must exert a bending moment upon all parts of the rib, as well as a direct thrust, which moment is too important to be neglected. It being recollected that the condition $\Sigma E F . D E = 0$ denoted that the change of span equalled zero, it will be sufficient in this case to still make it zero, when we have added or subtracted a quantity proportional to $2 t e c$.

72. **Change of Span influenced by Material and Cross-section of Arch.** — The bending moment M at any point has been demonstrated, § 4, to be equal to the product of H from the stress diagram multiplied by the vertical ordinate from that point to the equilibrium polygon. Then it was shown, § 18, that, if all these ordinates were summed up, that is, if w $\Sigma E F$ between two points, this sum would be *proportional* to the change of inclination between those two points; but it was not stated that this quantity was *equal* to the change of inclination, for neither the material nor the form of cross-section of the rib was taken into account. As the amount of flexure was stated, in Part II., "Bridges," §§ 85 and 86, to vary inversely as the modulus of elasticity and the moment of inertia, we must write $\frac{\Sigma M}{\mathbf{E I}}$ or $\frac{H . \Sigma E F}{\mathbf{E I}}$ to obtain a quantity which shall *equal* the change of inclination. The same thing is true of the expressions for deflection and change of span. When, however, the summation is made from one abutment to the other, and then put equal to zero, if **E** and **I** are constant, as well as H, it must be true that $\Sigma E F = 0$, as heretofore stated; and likewise of the other equations. Now **E** is constant, as the material of the rib is the same throughout; and since the parabolic rib, of cross-section varying with the secant of the inclination of the rib to the horizon, has been demonstrated, § 36, to deflect vertically like a straight beam of uniform section equal to that of the rib at the crown, **I** is likewise constant in these formulæ,

nd represents the moment of inertia of the section at the rown. In short, where one quantity is directly proportional to nother, if one is equal to zero, the other is also; consequently e can deal with areas, area moments, &c., as if they were the changes of inclination, deflections, &c., themselves.

73. **Formula for H from Change of Temperature.** — But now we wish to introduce the distance $2tec$, the change of span which would occur from change of temperature, were it unchecked. As this is an absolute and not a proportional uantity, we must divide our original quantity for change of pan, § 7, by **EI**. We shall, therefore, have for the new condition,

$$\frac{H_t \, . \, \Sigma \, EF \, . \, DE}{\mathbf{EI}} \pm 2tec = 0,$$

where H_t is used to signify the horizontal force (thrust or tension) which is occasioned by the change of temperature; or, if we clear of fractions, we get the more convenient expression

$$H_t \, . \, \Sigma \, EF \, . \, DE \pm 2\mathbf{EI}tec = 0.$$

A rise of temperature will make H a thrust or positive, while a fall of temperature will make H a tension or negative. The double sign is not needed in the above equation if the sign is contained in the symbol t, that is, if t is negative for a diminution of temperature below the one at which the rib is constructed or laid out. The bending moments exerted on the rib will be of the contrary kind when H_t is minus, while the ordinates are unchanged.

74. **Application to Parabolic Rib, Hinged at Ends.** — To take up first the case of the parabolic rib hinged at ends. The amount of H_t is to be determined. As there can be no bending moment at either abutment, and H_t at each abutment is the only applied force, the equilibrium polygon or line of thrust, Fig. 16, must be in the line joining the two springings. The bending moment at any point will, therefore, be equal to the ordinate to the rib at that point, multiplied by the desired value of H_t. The expression $\Sigma \, EF \, . \, DE$ therefore becomes for

this case $\Sigma D E^2$; and we have, transposing the second term of the equation of the previous section,

$$H_t \cdot \Sigma D E^2 = 2 E I t e c.$$

The value of $\Sigma D E^2$ was shown in § 39 (2.), to be $\frac{16}{15} k^2 c$; therefore, substituting and transposing, we see that

$$H_t = \frac{15}{8} \cdot \frac{t e E I}{k^2},$$

a value which is independent of the span.

The maximum bending moment, which occurs at the middle of the span, where the ordinate will be k, is

$$M \text{ (max.)} = \frac{15}{8} \cdot \frac{t e E I}{k}.$$

The ordinates at all the usual points of division will be the values of z, used repeatedly before; and, by multiplying H_t by these several values of z, the bending moments at all are obtained for a given change of temperature t. An additional line can be placed below the table of M to contain these quantities, so as to have them convenient for use. All of these moments will be positive for a fall of temperature below, and negative for a rise above, that at which the rib was designed. The worst effect of either change must be provided for.

75. Formula for Change of Span deduced analytically. — If one likes to prove this value for change of span analytically, he may proceed as follows: Let any ordinate to the arch be denoted by y, and the abscissa measured horizontally from one abutment by x. Then, if v = the vertical deflection ordinate, that is, the deflection of any point from its original position, we may write the usual equations for curvature, slope, and deflection of beams, recollecting that this arch acts like a beam of uniform section in deflecting vertically,

$$\frac{d^2 v}{d x^2} = \frac{M}{E I}; \quad \frac{d v}{d x} = \int \frac{M}{E I} \, d x; \text{ and } v = \int\int \frac{M}{E I} \, d x^2.$$

Now $M = H y = H \frac{k}{c^2} (2 c x - x^2)$; therefore

$$\frac{d v}{d x} = \frac{H}{E I} \cdot \frac{k}{c^2} \int (2 c x - x^2) \, d x = \frac{H}{E I} \cdot \frac{k}{c^2} \left(c x^2 - \frac{x^3}{3} + C \right).$$

or $x = c$; therefore $C = -\frac{2}{3}c^3$. Then

$$\frac{dv}{dx} = \frac{H}{EI} \cdot \frac{k}{c^2}\left(cx^2 - \tfrac{1}{3}x^3 - \tfrac{2}{3}c^3\right). \quad (a.)$$

horizontal displacement of any point, the infinitesimal horizontal ment du, due to the movement of the portion of arc ds, will give, be seen to the right of Fig. 16,

$$du : dv = dy : dx.$$

se $y = \frac{k}{c^2}(2cx - x^2)$, $dy = \frac{k}{c^2}(2c - 2x)\,dx$, and we have

$$du = \frac{2k}{c^2}(c - x)\,dv.$$

Substitute the value of dv from (a.), and it becomes

$$du = \frac{H}{EI} \cdot \frac{2k^2}{c^4}\left(c^2x^2 - \tfrac{4}{3}cx^3 - \tfrac{2}{3}c^4 + \tfrac{1}{3}x^4 + \tfrac{2}{3}c^3x\right)dx.$$

If this equation is integrated between the limits 0 and $2c$, we obtain $u = -\frac{H}{EI} \cdot \frac{16}{15}k^2c$, which will be seen to correspond with the value of $2tec$ in the preceding section.

76. **Application to Fixed Parabolic Rib.** — If we turn next to the rib with fixed ends, it will be manifest, that, since there will be bending moments at the springings, the line which corresponds to the equilibrium polygon and limits the ordinates for bending moments cannot now pass through those points. As the resistance to expansion or contraction is the only cause of those moments, the two abutment moments will be equal, and the line will be horizontal. In order also to satisfy the condition that the change of inclination at the abutments shall equal zero, or, as expressed in § 18, $\Sigma EF = 0$, the horizontal line must be so drawn as to make the areas within and without the arch equal to one another, which will occur when the line is drawn at a height of $\frac{2}{3}k$ above the springing, as seen in Fig. 17. To prove the equality of areas it is only necessary to recall the fact that the area of a parabolic segment equals two-thirds of the enclosing rectangle. The area included within the

whole arch will therefore be $\frac{2}{3}k \,.\, 2c = \frac{4}{3}kc$. The rectangle of height $\frac{2}{3}k$ has the same area. Therefore the portions of the arch area and of the rectangle which do not coincide must be equal to one another. The third condition, of § 19, that $\Sigma\, EF \,.\, DB = 0$, or the equality of area moments, is also satisfied by this construction; for the rectangle multiplied by the half span, which is the distance of its centre of gravity from one abutment, is equal to the area included by the whole arch multiplied by the same distance.

To deduce in this case the value of H_t: as before,

$$H_t \,.\, \Sigma EF \,.\, DE \pm 2EItec = 0. \quad (1.)$$

From what has just been stated,

$$\Sigma EF \,.\, DE = \Sigma (DE - \tfrac{2}{3}k)\, DE = \Sigma DE^2 - \tfrac{2}{3}k \,.\, \Sigma DE. \quad (2.)$$

The first term, as before, amounts to $\frac{16}{15}k^2c$; since ΣDE = enclosed by the arch, $= \frac{4}{3}kc$, the second term is $\frac{8}{9}k^2c$; ı fore

$$H_t \,.\, \tfrac{8}{45}k^2c = 2EItec, \quad \text{or} \quad H_t = \tfrac{45}{4}\frac{teEI}{k^2}.$$

The bending moment at the crown will therefore be

$$M = H_t \,.\, \tfrac{1}{3}k = \tfrac{15}{4} \,.\, \frac{teEI}{k},$$

and at the springing,

$$M = H_t \,.\, \tfrac{2}{3}k = \tfrac{15}{2} \,.\, \frac{teEI}{k},$$

or double the former amount, but of the opposite kind. Whether the bending moment at either point is positive or negative, depends upon whether H_t is tension or compression. These moments also can be conveniently added to the proper table for M, as explained for the first case.

77. **Comparison of Arches under Change of Temperature.** — The bending moments for temperature, in both the arch with hinged ends and that with fixed ends, will vary like those of a beam uniformly loaded, and either simply supported or fixed at the ends. Part II., "Bridges," §§ 95, 99.

It may be well to notice the comparative straining effect of the same change of temperature in the two classes of parabolic arches, for ribs of the same rise. H_t is six times as great when the arch is fixed as when it is hinged at the ends, and the direct stress in the ribs will therefore vary in the same proportion. The maximum moment, at the springing, for the rib with fixed ends, is four times as great as at the crown of the rib with hinged ends, and of the opposite kind; while the value of M at the two crowns is as two to one against the rib with fixed ends.

78. **Shear from Change of Temperature.** — The shear on a right section can be shown by the accompanying Fig. 18. If *a b* represents the amount of H caused by a change of temperature, we may draw *a d* and *b c* parallel to the upper and lower flange at any right section S of the rib, when *e a* will be the value of the direct stress at the section, one-half in each flange, and *b e* will be the shear.* The bending moment will have any magnitude, depending upon the length of the ordinate from the equilibrium line to the point on the centre line of the arch where this section is taken. As *a e* and *g b* are parallel, the perpendicular distance *b e*, = *c d*, between them is constant, so that *f d* may be taken, for our purpose, to represent the stress in one chord, and *g c* that in the other due to bending moment, the resultant stresses being *a d* and *c b*, while the shear on the right of a right section of the web will be *d c*. Since the resultant stress at any section must be H, the directions of the forces, shown by the arrows, in this closed polygon, are at once fixed. As the inclination of the arch changes, the value of *c d* will change, being zero at the crown and a maximum at the springings. The arrows denote the case where H is a thrust. The bending moment will be negative, if the rib is hinged at the ends, the bottom chord will be compressed, the top chord will have a force exerted upon it amounting to the difference between the direct thrust and the tension due to the moment, and consequently *c b* will be the stress exerted by the top chord against the right side of the cross-section in the accompanying sketch.

79. **Diagram for Vertical Shear.** — Let us suppose a fall of

* In Fig 18, the point *f* should bisect *e a*.

temperature to take place; the rib will have a tendency to come down at the crown. We recall the fact that a uniform load has a parabola for its equilibrium curve, and a load of the proper intensity on any parabolic arch will produce the value of H which is now supposed to exist. It is evident, then, as is also shown by the sign of M, that the rib may be imagined to be loaded uniformly horizontally with a weight sufficient to produce this deflection or these values of M. This imaginary weight will be just sufficient at all points to balance the component of an opposite kind which is required in combination with the value of H_t (in this case a horizontal tension), in order to give a resultant stress in the direction of the tangent to the rib. And, further, if this weight were not just sufficient to balance the above component, a remainder, of one sign or the other, would be found at the abutments, as a vertical component of the reaction there; but we know that no such vertical component exists. If a bent spring is placed with its two ends on a horizontal line, and compression or tension is applied in that line, no vertical force is needed for equilibrium. As the uniform weight was entirely imaginary, the vertical components must be supplied by the web and flanges, and hence we conclude that the diagram for *vertical* shear in the arch affected by a change of temperature, will be that of an ordinary truss, supported or fixed at its two ends, and carrying a complete uniform load, and that the normal component will be carried by the web. For a fall of temperature, therefore, the shear on a vertical section will be of the same kind as, and, for a rise of temperature, will be of the opposite kind to, that produced by a load on a truss with horizontal chords.

CHAPTER VI.

CIRCULAR RIB WITH HINGED ENDS.

80. **Circular Rib to be of Uniform Section.** — Passing next to the consideration of the arch whose curve is the arc of a circle, we shall assume that the rib is of uniform section, and not, as before, of increasing breadth from the crown to the ɪging. As the rib is of uniform section, it can no longer be compared to a horizontal beam, as regards its change of inclination and deflection under bending moments, and the length along the arch, instead of its projection on a horizontal line, must be used in spacing off and in summing up the usual quantities; that is, the sum of the changes of inclination between any two points will be made up from the change of inclination at each successive point *along the rib.* We must therefore use ds for dx in our integration, where s denotes the length of an arc; and polar co-ordinates will, in the more complex cases, be used in place of rectangular ones. In spacing off the rib for equal divisions, or for summing the ordinates arithmetically, the measurements will be made along the curve, and each division will subtend the same angle at the centre of the circle.

We stated, it will be remembered, that a segmental arch of the circular type, if the rise did not exceed one-tenth of the span, might, without serious error, be treated as if it were parabolic. In discussing circular arches, there will be so many points similar to those we have already explained, that we shall not go into much detail on some points, but leave the reader to make the extended application as examples come up in his own practice.

81. **Experimental Verification.** — The values to be obtained for y_0, for a rib of uniform section, curved to the arc of a circle, and hinged or free to turn at the ends, can be readily verified or illustrated experimentally as follows: — Take a piece of moderately stiff iron wire, and bend it accurately into the desired shape, A C B, Fig. 19; suspend the wire from a horizontal bar E F by means of strings fastened at A and B, and then attach a weight at any point C. It will be convenient to stretch a thread from A to B, which, as the span is to be unchanged, will not interfere with the reactions. If the point E is now moved horizontally, the length of the string E A being at the same time changed, the line A B can be brought parallel with E F, as can be readily ascertained with a scale. Then E A and F B prolonged will meet at D on C D, and D G will equal y_0. E A and F B will actually intersect on the vertical through the centre of gravity of the wire and weight combined; but if the weight of the wire is as small as is consistent with stiffness, while the weight at C is large in comparison, the centre of gravity will practically be in C D. If A B becomes slack, it shows that E and F are not sufficiently far apart. By fastening two long threads independently to E and F, the lines E A and F B can be easily prolonged to an intersection.

82. **Semicircular Arch with Hinged Ends; Value of y_0.** — If the rib with hinged ends is first taken up for discussion, the value of y_0 for a load at any point on a semicircular arch is easily obtained by a simple device. Recurring again to the usual formula in its modified form, we must satisfy the condition

$$\Sigma \, DE^2 = \Sigma \, DE \cdot DF.$$

If we let D E, Fig. 20, $= z$; D F $= y$; A D $= x$; and represent a small portion of arc by ds, this equation becomes, for the entire semicircle,

$$\int_0^\pi z^2 \, ds = \int_0^\pi yz \, ds.$$

If we draw a radius from any point E of the rib to the centre O, and also draw the infinitesimal triangle whose sides are ds, dx, and dz, we shall have, from similarity of triangles,

$$r : z = ds : dx, \text{ or } z\,ds = r\,dx;$$

substituting this value in the above equation, we get

$$r \int_0^{2c} z\,dx = r \int_0^{2c} y\,dx.$$

The integral of $z\,dx$ between the given limits is the area of the semicircle, while that of $y\,dx$ is the triangle A C B. Substitute the value $\frac{1}{2}\pi r^2$ for the former, and $r\,y_0$ for the latter, and we obtain

$$\tfrac{1}{2}\pi r^3 = r^2 y_0; \text{ or } y_0 = \tfrac{1}{2}\pi r = 1.5708\,r.$$

The ordinate y_0, for a load at any point, on a semicircular rib with hinged ends, is therefore a constant quantity, equal to the length of the half rib. If we draw a horizontal line at this ight above the springing, it will contain the vertices of all tne equilibrium polygons for single loads.

83. **Segmental Arch; Value of y_0.** — If the arch is segmental, that is, less than a semicircle, we shall use the following notation: Let the angle N O B, Fig. 21, subtended at the centre of the circle by the half arch, be denoted by β; the angle N O I, from the crown to the point where the weight is placed, be denoted by α; and the angle N O E, from the crown to any point where the ordinates D E and E F are measured, be θ. The radius of the arch $= r$. If, then, A C B is the desired curve of equilibrium, C K $= y_0$. The value of this ordinate will be proved to be

$$y_0 = r \frac{(\sin^2\beta - \sin^2\alpha)\left(\beta\,\dfrac{1 + 2\cos^2\beta}{\sin\beta} - 3\cos\beta\right)}{(\sin^2\beta - \sin^2\alpha) + 2\cos\beta\,(\alpha\sin\alpha + \cos\alpha - \beta\sin\beta - \cos\beta)}.$$

If the arch is a semicircle, $\beta = 90° = \tfrac{1}{2}\pi$, and this value reduces to $y_0 = \tfrac{1}{2}\pi r$, as previously obtained.

The work of computing y_0 for different values of α is not great; as, for a given arch, β is constant, and the second factor of the numerator is a constant quantity. Since a segmental arch may subtend any angle, it is not worth while to go into the computation here of values of y_0 for a given value of β; but, as examples of y_0, we will give

If $\beta = 45°$	and $\alpha = 0°$,	then $y_0 = .39\,r$	nearly.
" 45°	" 30°,	" .42 r	"
" 60°	" 30°,	" .71 r	"

All that one needs for the calculation from this formula is an ordinary table of natural sines and cosines. The angles or arcs β and α are to be expressed in lengths of arc, which subtend the given number of degrees, to radius unity. The arc for one degree being $\frac{\pi}{180}$, or 0.017453, any other arc will be obtained by multiplying this quantity by the number of degrees wh the arc subtends, minutes being expressed as a decimal part of a degree.

84. **Proof.** — From Fig. 21 we have $DE = r(\cos\theta - \cos\beta)$.

$$DF : CK = AD : AK = r(\sin\beta + \sin\theta) : r(\sin\beta + \sin\alpha)$$

on the left of K, or $DF = \frac{\sin\beta + \sin\theta}{\sin\beta + \sin\alpha} y_0$;

on the right of K, $DF = \frac{\sin\beta - \sin\theta}{\sin\beta - \sin\alpha} y_0$.

Substituting these values in the usual equation, § 39, $\Sigma DE^2 = \Sigma DE \, . \, DF$, we obtain for the first member of the equation, remembering to use $ds = r\,d\theta$ in place of dx, and considering angles to the left of ON as negative,

$$r^3 \int_{-\beta}^{+\beta} (\cos\theta - \cos\beta)^2 d\theta = r^3 \int_{-\beta}^{+\beta} (\cos^2\theta - 2\cos\beta\cos\theta + \cos^2\beta)\, d\theta^*$$

$$= r^3(\beta + 2\beta\cos^2\beta - 3\sin\beta\cos\beta). \quad (a.)$$

For the integral of the second member between α and $-\beta$ we have

$$\frac{r^2 y_0}{\sin\beta + \sin\alpha} \int_{-\beta}^{\alpha} (\sin\beta\cos\theta + \sin\theta\cos\theta - \sin\beta\cos\beta - \cos\beta\sin\theta)\, d\theta\, \dagger$$

---

* $\int \cos^2\theta\, d\theta = \frac{1}{2}(\theta + \sin\theta\cos\theta)$; $\cos -\beta = \cos\beta$; $\sin -\beta = -\sin\beta$.

† $\int \sin\theta\cos\theta\, d\theta = -\frac{1}{2}\cos^2\theta$.

$$= \frac{r^2 y_0}{\sin\beta + \sin\alpha}(\sin\alpha\sin\beta - \tfrac{1}{2}\cos^2\alpha - \alpha\sin\beta\cos\beta$$
$$+ \cos\alpha\cos\beta + \sin^2\beta - \tfrac{1}{2}\cos^2\beta - \beta\sin\beta\cos\beta).$$

—ise for the integral of the second member between α and $+\beta$ we have

$$\frac{r^2 y_0}{-\sin\alpha}\int_\alpha^\beta (\sin\beta\cos\theta - \sin\theta\cos\theta - \sin\beta\cos\beta + \cos\beta\sin\theta)\,d\theta$$

$$= \frac{r^2 y_0}{\sin\beta - \sin\alpha}(\sin^2\beta - \tfrac{1}{2}\cos^2\beta - \beta\sin\beta\cos\beta - \sin\alpha\sin\beta$$
$$- \tfrac{1}{2}\cos^2\alpha + \alpha\sin\beta\cos\beta + \cos\alpha\cos\beta).$$

two quantities are to be reduced to a common denominator, added ...er and equated with the first member (a.). Upon making simple ellations, dividing through by $\sin\beta$, and factoring, we get the form of iven in the last section.

85. **Formula for H; Value of Ordinates.**—When the value, of y_0 is computed, we can readily draw the stress diagram of Fig. 21, and scale the value of H; or the formula proved before, § 40, may be applied here, and is easily converted into the third form,

$$H = \frac{W}{y_0} \cdot \frac{c^2 - b^2}{2c} = W\frac{AK \cdot KB}{CK \cdot AB} = \frac{r(\sin^2\beta - \sin^2\alpha)}{y_0 \cdot 2\sin\beta} W. \quad (1.)$$

If calculations have already been made for y_0, the quantities desired for this formula are at hand.

Then the ordinate at each point of division, by which H is to be multiplied to give M for that point, will be, from § 84, if θ is the angle between the two radii from the crown and the point E,

$$EF = DF - DE = y_0\frac{\sin\beta \pm \sin\theta}{\sin\beta \pm \sin\alpha} - r(\cos\theta - \cos\beta). \quad (2.)$$

The plus sign is to be used for points between the weight and the farther abutment, and the minus sign between the weight and the nearer abutment. We must remember, however, that, if θ is measured from the crown to the right as the positive direction, all angles θ on the left of the crown will be negative, and their *sines* will be minus. If E F is plus, it gives a positive bending moment, tending to make the arch less convex, and *vice versa*.

86. **Numerical Computation of M.**—In any practical case *we should much prefer*, as more easy and sufficiently accurate, *to scale all of these quantities from a good-sized diagram;* but it may be well to compute one set

of values of M as an example, for fear the signs may give some readers trouble. Taking the case of Fig. 22, let $\beta = 45°$ and $\alpha = 20°$. Then the arc $\beta = .7854$ and $\alpha = .3491$; $\sin\beta = \cos\beta = .7071$; $\sin\alpha = .3420$, $\cos\alpha = .9397$. These values, substituted in the equation of § 83, give

$$y_0 = r\frac{(.5 - .1170)\left(.7854\dfrac{2}{.7071} - 2.1213\right)}{.5 - .1170 + 1.4142\,(.1194 + .9397 - .5554 - .7071)} = \frac{.0384}{.0954}r = .403\,r;$$

(1.), § 85, will then become

$$H = \frac{(.5 - .1170)\,r}{1.4142 \times .403\,r}W = \frac{.383}{.570}W = .672\,W.$$

$$\text{Sin}\,\beta + \sin\alpha = 1.0491;\quad \sin\beta - \sin\alpha = .3651;$$

$$\frac{y_0}{\sin\beta + \sin\alpha} = \frac{.403\,r}{1.0491} = .384\,r;\quad \frac{y_0}{\sin\beta - \sin\alpha} = \frac{.403\,r}{.3651} = 1.104\,r.$$

VALUES OF M.

W.

θ	$-40°$	$-30°$	$-20°$	$-10°$	$0°$	$10°$	$20°$	$30°$	$40°$	
$\sin\beta =$	.7071								.7071	$\sin\beta$
$+\sin\theta$	—.6428	—.5	—.3420	—.1736	0	+.1736	+.3420	.5	.6428	$-\sin\theta$
Mult. by	.0643	.2071	.3651	.5335	.7071	.8807	1.0491	.2071	.0643	Mult. by
.384 =	.0247	.0795	.1402	.2049	.2715	.3382	.4029	.2286	.0710	1.104 =
$-\cos\theta$	.7660	.8660	.9397	.9848	1.0	.9848	.9397	.8660	.7660	$-\cos\theta$
	—.7413	—.7865	—.7995	—.7799	—.7285	—.6466	—.5368	—.6374	—.6950	
$+\cos\beta$	.7071								.7071	$+\cos\beta$
	—.0342	—.0794	—.0924	—.0728	—.0214	+.0605	+.1703	+.0697	+.0121	
$\times .672$ W	—.0230	—.0534	—.0621	—.0489	—.0144	+.0407	+.1144	+.0468	+.0081	r W = M

87. **Shear at any Right Section.** — Suppose that the rib of Fig. 22 carries a single weight under the point C, and that the curve of equilibrium is A C B. If 012 is the stress diagram, 2–3 will be the vertical component of the reaction at A, and 3–1 that at B. To find the shear on a right section near A, as at E, lay off 2–3, or P_1 in Fig. 23, and draw H so that the arrows may follow one another; then from 0 draw a line 0–4 parallel to the tangent at E; the perpendicular distance 4–2 will be the

shear in the web. For we see by the direction of the arrows that these forces last drawn balance P_1 and H, and, as in Fig. 18, no matter how much the bending moment, and hence the flange stress, may be, the perpendicular distance 4–2 is unchanged. The line 0–4 will be the magnitude of the direct thrust. Both of these forces are given on the right of the section, and this shear is therefore negative. In the same way, for the point E near B, draw $1–3 = -P_2$ and $3–0 = H$; draw 0–8 parallel to the tangent at E; 8–1, perpendicular to it, will be the shear on the right of the section, again negative, and 0–8 will be the direct thrust. It is noticeable that the normal shear in the web near the left abutment is opposite in sign to P_1, while near the right abutment it agrees in sign with P_2. For the kind of brace needed, see Fig. 10. It is evident that these figures may at once be drawn on the stress diagram, where 0–4 and 4–2 are already sketched. Such a way will answer well for a few points on a large figure, especially if we have applied such loads as give the maximum shear at any particular point. If, however, we desire to see the variation of the shear across the span, we may draw a different diagram.

88. **Shear Diagram.** — As the tangent is perpendicular to the radius at the point of contact, we may at once see that the angles marked θ in Fig. 23 correspond with the angle θ made by the radius to the crown and that to the point E. Hence we get a value for the normal shear, $P \cos \theta - H \sin \theta$. As the point E is distant horizontally from the middle of the span an amount $r \sin \theta$, the last term of this expression for shear varies directly as the distance from the centre; and if we draw 3–7, in the stress diagram of Fig. 22, parallel to the radius at A, cutting 0–6 which is parallel to the tangent at A, 3–7 will be $H \sin \theta$ for A, and may be laid off at $a\,w$ and $b\,r$ of Fig. 23. The vertical ordinate $e\,d$ will then represent $H \sin \theta$ at any point. P_1 is laid off at $c\,l$, and P_2 at $c\,m$; with c as centre, and these two distances as radii, draw the dotted arcs seen in the figure; lay off several angles θ at c, as, for instance, $l\,c\,g$ and $m\,c\,n$ for the points E; project g and n horizontally to f under the respective points E;

df will be $P \cos\theta$, and from several similarly located points the curves *slt* and *vfr* are found. Then at any point the *vertical* distance $df - ed$ or *ef* will be the *normal* shear in the web on the left of the section, positive if above the inclined line, negative if below it.

From the formula $P\cos\theta - H\sin\theta$, a table of shears may be easily computed for any given arch. $P\sin\theta + H\cos\theta$ will give the direct thrust.

89. **Distribution of Load to produce Equilibrium.** — A series of lines drawn in the stress diagram from 0, parallel to the tangents at a number of equidistant points in a circular rib, will cut off such portions of the load line as represent the loads necessary to make the successive sides of the equilibrium polygon parallel to these tangents, or, in short, coincident with the rib. But the lines radiating from 0 will successively inter increasing lengths of load line. Hence the load which will keep a circular arch in equilibrium must increase in intensity per horizontal foot from the crown to the springing, and must become infinite at the springing of a semicircular arch. Hence it follows that no amount and distribution of vertical load can make a semicircular arch a true equilibrium curve, that is, one which has no bending moment at any point. In fact, no curve which starts vertically from the abutment can be an equilibrium curve under vertical loads. This may be seen in a more simple manner if we consider that no arrangement of weights will cause a cord, attached at two points, to hang in a funicular polygon whose first side is vertical.

90. **Effect of Change of Temperature.** — The horizontal thrust or tension, due to a change of temperature, in a circular rib hinged at the ends, is found by a similar method to that pursued for the parabolic rib. Referring, to avoid repetition, to what was said at that time, §§ 71–73, the equation may be written, as given in § 74,

$$H_t \cdot \Sigma DE^2 = \pm 2EI \cdot tec.$$

Fig. 16 will answer for this case, if we imagine the arc to be

As we saw, in § 82, that $\Sigma \mathrm{D\,E}^2$ for a semicircular πr^3, a substitution in the above equation gives at

$$H_t = \pm \frac{4\,\mathrm{E\,I}\,.\,tec}{\pi r^3} = \pm\ 1.264\ \frac{\mathrm{E\,I}\,te}{r^2}$$

emicircular rib. The bending moment at the crown, ; is a maximum, will be

$$M\ (\text{max.}) = \frac{4\,\mathrm{E\,I}\,te}{\pi r}.$$

ıe arch is less than a semicircle, (*a.*), § 84, gives

$$\Sigma \mathrm{D\,E}^2 = r^3\,(\beta + 2\,\beta \cos^2 \beta - 3 \sin \beta \cos \beta),$$

and $c = r \sin \beta$; therefore, substituting, we obtain

$$H_t = \pm \frac{2\,\mathrm{E\,I}\,te \sin \beta}{r^2\,(\beta + 2\,\beta \cos^2 \beta - 3 \sin \beta \cos \beta)},$$

and the bending moment at the crown will be

$$M\ (\text{max.}) = \frac{2\,\mathrm{E\,I}\,te \sin \beta\,(1 - \cos \beta)}{r\,(\beta + 2\,\beta \cos^2 \beta - 3 \sin \beta \cos \beta)}.$$

91. **Shear from Change of Temperature.** — If a load of the proper amount and distribution were imposed on the rib to place it entirely in equilibrium, and cause it to exert against the abutments the desired value of H due to temperature, such a load would supply the amount of shear needed at each section, and, when the load is absent, the bracing must supply such shear. The line *wecer* of the shear diagram of Fig. 23 will therefore limit the ordinates for shear at right sections of the web under changes of temperature, when 0–3 is the amount of H_t. A reference to § 78 and § 87 will aid the reader in recalling these points.

CHAPTER VII.

CIRCULAR RIB WITH FIXED ENDS.

92. Values of Equations of Condition. — When th cular rib is fixed at the ends, we apply the three equatio condition which were developed in §§ 17–19, summing up ordinates, however, along the arch, as has just been done in th preceding case, in place of the horizontal line. When the arch is a complete semicircle, or, as it is often called, a complete arch, as distinguished from a segmental one, the value of y_0, y_1, and y_2 may be obtained by a device similar to the one employed in § 82. The equation to satisfy the first condition is easily derived, but the two others present more difficulty; it is therefore not expedient to take up the semicircle as a special case, but rather to work out the general equations, and make the necessary substitutions.

In the arch of Fig. 24, let A N $= y_1$, C K $= y_0$, and B R $= y_2$; M O B = M O A $= \beta$, M O I $= \alpha$, and M O E, to any point E, $= \theta$, angles to the right of M being positive. The notation agrees with that just used. Then it may be proved that the three equations of condition will reduce to

$$\sin\beta\, y_0 + \tfrac{1}{2}(\sin\beta + \sin\alpha)\, y_1 + \tfrac{1}{2}(\sin\beta - \sin\alpha)\, y_2 = (\beta - \sin\beta\cos\beta)\, r; \quad (1.)$$

$$\begin{aligned} &- \sin\beta(\cos\alpha - \cos\beta + \alpha\sin\alpha - \beta\sin\beta)\, y_0 \\ &+ \tfrac{1}{2}(\sin\beta - \sin\alpha)(\cos\alpha - \cos\beta + \alpha\sin\alpha + \beta\sin\alpha)\, y_1 \\ &+ \tfrac{1}{2}(\sin\beta + \sin\alpha)(\cos\alpha - \cos\beta + \alpha\sin\alpha - \beta\sin\alpha)\, y_2 \\ &= (\sin\beta - \beta\cos\beta)(\sin^2\beta - \sin^2\alpha)\, r; \quad (2.) \end{aligned}$$

$$[(\beta - \cos\beta \sin\beta)\sin\alpha - (\alpha + \sin\alpha\cos\alpha - 2\sin\alpha\cos\beta)\sin\beta]\, y_0$$
$$+ \tfrac{1}{2}(\sin\beta - \sin\alpha)(\alpha + \sin\alpha\cos\alpha + \beta - \sin\beta\cos\beta - 2\sin\alpha\cos\beta)\, y_1$$
$$+ \tfrac{1}{2}(\sin\beta + \sin\alpha)(\alpha + \sin\alpha\cos\alpha - \beta + \sin\beta\cos\beta - 2\sin\alpha\cos\beta)\, y_2 = 0. \quad (3.)$$

It will be easier to solve the numerical equations after the values of α and β, with their sines and cosines, are introduced, than to deduce independent values of y_1, &c., at present. They may be written more briefly, for convenience in substitution, if

$$\sin\beta - \sin\alpha = a;\ \sin\beta + \sin\alpha = b;\ \alpha + \sin\alpha\cos\alpha - 2\sin\alpha\cos\beta = c;$$
$$\beta - \sin\beta\cos\beta = d;\ \cos\alpha - \cos\beta + \alpha\sin\alpha = e;$$

$$\sin\beta\, y_0 + \tfrac{1}{2} b\, y_1 + \tfrac{1}{2} a\, y_2 = d r. \quad (4.)$$

$$-(e - \beta\sin\beta)\sin\beta\, y_0 + \tfrac{1}{2} a (e + \beta\sin\alpha)\, y_1 + \tfrac{1}{2} b (e - \beta\sin\alpha)\, y_2$$
$$= a b (\sin\beta - \beta\cos\beta)\, r. \quad (5.)$$

$$(d\sin\alpha - c\sin\beta)\, y_0 + \tfrac{1}{2} a (c + d)\, y_1 + \tfrac{1}{2} b (c - d)\, y_2 = 0. \quad (6.)$$

cial Values for Semicircular Rib. — If the arch is :cle, $\beta = \frac{1}{2}\pi$; $\sin\beta = 1$; $\cos\beta = 0$; and the three equa- . of the last section reduce to

$$y_0 + \tfrac{1}{2}(1 + \sin\alpha)\, y_1 + \tfrac{1}{2}(1 - \sin\alpha)\, y_2 = \tfrac{1}{2}\pi r; \quad (1.)$$

$$(\tfrac{1}{2}\pi - \cos\alpha - \alpha\sin\alpha)\, y_0 + \tfrac{1}{2}(1 - \sin\alpha)(\cos\alpha + \alpha\sin\alpha + \tfrac{1}{2}\pi\sin\alpha)\, y_1$$
$$+ \tfrac{1}{2}(1 + \sin\alpha)(\cos\alpha + \alpha\sin\alpha - \tfrac{1}{2}\pi\sin\alpha)\, y_2 = (1 - \sin^2\alpha)\, r: \quad (2.)$$

$$(\tfrac{1}{2}\pi\sin\alpha - \alpha - \sin\alpha\cos\alpha)\, y_0 + \tfrac{1}{2}(1 - \sin\alpha)(\alpha + \sin\alpha\cos\alpha + \tfrac{1}{2}\pi)\, y_1$$
$$+ \tfrac{1}{2}(1 + \sin\alpha)(\alpha + \sin\alpha\cos\alpha - \tfrac{1}{2}\pi)\, y_2 = 0. \quad (3.)$$

If equation (1.) is multiplied by α, equation (3.) may be added to it, and then (2.) may be multiplied by $\sin\alpha$, and subtracted from their sum, when there will result

$$(\alpha + \tfrac{1}{4}\pi - \tfrac{1}{4}\pi\sin\alpha)\, y_1 + (\alpha - \tfrac{1}{4}\pi - \tfrac{1}{4}\pi\sin\alpha)\, y_2 = (\tfrac{1}{2}\pi\alpha - \sin\alpha)\, r. \quad (4.)$$

If (1.) is multiplied by $\frac{1}{2}\pi - \cos\alpha - \alpha\sin\alpha$, and equation (2.) is subtracted from it, we shall get, upon dividing by the common coefficient of y_1 and y_2,

$$\tfrac{1}{2}(y_1 + y_2) = \frac{\frac{1}{2}\pi(\frac{1}{2}\pi - \cos\alpha - \alpha\sin\alpha) - \cos^2\alpha}{\frac{1}{2}\pi - 2\cos\alpha - 2\alpha\sin\alpha + \frac{1}{2}\pi\sin^2\alpha}\, r,$$

which, if the quantity in the parentheses be represented by g, may be written,

$$\tfrac{1}{2}(y_1 + y_2) = \frac{\frac{1}{2}\pi g - \cos^2\alpha}{2g - \frac{1}{2}\pi\cos^2\alpha}\, r. \quad (5.)$$

Upon multiplying this equation by $2\alpha - \frac{1}{2}\pi \sin \alpha$, and subtracting it from (4.), we obtain, by factoring the second member,

$$\frac{1}{2}(y_1 - y_2) = -\frac{\left(\frac{4}{\pi} - \frac{\pi}{2}\right)(a \cos^2 a - g \sin a)}{2g - \frac{1}{2}\pi \cos^2 a} r. \quad (6.)$$

The sum of (5.) and (6.) will give y_1; their difference will give y_2; and these values, inserted in (1.), will readily give us y_0.

94. **First Equation of Condition.** — Many of the following expressions are similar to those of § 84, and a remembrance of the relation between y_1 and y_2 will, in a measure, prevent the ensuing work from seeming so involved as it otherwise may appear. Generally, coefficients of y_1 and y_2 will differ only in the signs of the terms which contain a and sine a. The first condition is

$$\Sigma \, \text{D E}^2 = \Sigma \, \text{D F} \cdot \text{D E}.$$

From § 84, we have

$$\Sigma \, \text{D E}^2 = r^3 (\beta + 2\beta \cos^2 \beta - 3 \sin \beta \cos \beta).$$

It will be seen, from Fig. 24, that $\text{D F} = \text{D L} + \text{L F} = y_1$ (or y_2) $+ \text{L F}$, D L in the sketch being negative on the right of K, and that, therefore, in place of the values of the section just referred to, we shall write

$$\text{D F} = y_1 + \frac{\sin \beta + \sin \theta}{\sin \beta + \sin a}(y_0 - y_1), \text{ on the left of K;}$$

$$\text{D F} = y_2 + \frac{\sin \beta - \sin \theta}{\sin \beta - \sin a}(y_0 - y_2), \text{ on the right of K.}$$

For the value of the second member of the above equation of condition between a and $-\beta$ we have then, since $\text{D E} = r(\cos \theta - \cos \beta)$,

$$r^2 \int_{-\beta}^{a} [y_1 (\cos \theta - \cos \beta) + \frac{y_0 - y_1}{\sin \beta + \sin a}(\sin \beta \cos \theta + \sin \theta \cos \theta - \sin \beta \cos \beta$$
$$- \cos \beta \sin \theta)]^* \, d\theta = r^2 [y_1 (\sin a - a \cos \beta + \sin \beta - \beta \cos \beta)$$
$$+ \frac{y_0 - y_1}{\sin \beta + \sin a}(\sin a \sin \beta - \tfrac{1}{2}\cos^2 a - a \sin \beta \cos \beta + \cos a \cos \beta$$
$$+ \sin^2 \beta - \tfrac{1}{2}\cos^2 \beta - \beta \sin \beta \cos \beta)].$$

Likewise, for the value of the second member between a and $+\beta$

* Compare § 84.

$$r^2\int_\alpha^\beta [y_2(\cos\theta - \cos\beta) + \frac{y_0 - y_2}{\sin\beta - \sin\alpha}(\sin\beta\cos\theta - \sin\theta\cos\theta - \sin\beta\cos\beta$$

$$+ \cos\beta\sin\theta)]^* d\theta = r^2[y_2(\sin\beta - \beta\cos\beta - \sin\alpha + \alpha\cos\beta)$$

$$+ \frac{y_0 - y_2}{\sin\beta - \sin\alpha}(\sin^2\beta - \tfrac{1}{2}\cos^2\beta - \beta\sin\beta\cos\beta - \sin\alpha\sin\beta - \tfrac{1}{2}\cos^2\alpha$$

$$+ \alpha\sin\beta\cos\beta + \cos\alpha\cos\beta)].$$

Equating the sum of these two quantities which make up the second member, with the first member, we obtain the first equation of condition, which, when cleared of fractions, becomes

$$y_0(2\sin^3\beta - \sin\beta\cos^2\beta - 2\beta\sin^2\beta\cos\beta - \cos^2\alpha\sin\beta + 2\cos\alpha\sin\beta\cos\beta$$
$$- 2\sin^2\alpha\sin\beta + 2\alpha\sin\alpha\sin\beta\cos\beta) + y_1(\tfrac{1}{2}\sin\beta\cos^2\beta - \sin^3\alpha$$
$$+ \alpha\sin^2\alpha\cos\beta + \beta\sin^2\alpha\cos\beta + \tfrac{1}{2}\cos^2\alpha\sin\beta - \cos\alpha\sin\beta\cos\beta$$
$$- \tfrac{1}{2}\sin\alpha\cos^2\alpha - \alpha\sin\alpha\sin\beta\cos\beta + \sin\alpha\cos\alpha\cos\beta + \sin\alpha\sin^2\beta$$
$$- \tfrac{1}{2}\sin\alpha\cos^2\beta - \beta\sin\alpha\sin\beta\cos\beta) + y_2(\tfrac{1}{2}\sin\beta\cos^2\beta + \sin^3\alpha$$
$$- \alpha\sin^2\alpha\cos\beta + \beta\sin^2\alpha\cos\beta + \tfrac{1}{2}\cos^2\alpha\sin\beta - \cos\alpha\sin\beta\cos\beta$$
$$+ \tfrac{1}{2}\sin\alpha\cos^2\alpha - \alpha\sin\alpha\sin\beta\cos\beta - \sin\alpha\cos\alpha\cos\beta - \sin\alpha\sin^2\beta$$
$$+ \tfrac{1}{2}\sin\alpha\cos^2\beta + \beta\sin\alpha\sin\beta\cos\beta) = r(\sin^2\beta - \sin^2\alpha)(\beta + 2\beta\cos^2\beta$$
$$- 3\sin\beta\cos\beta).$$

95. Second Equation of Condition. — The next condition to be satisfied is $\Sigma\,D\,E = \Sigma\,D\,F$, or, introducing the values of these quantities from the preceding section,

$$r^2\int_{-\beta}^{+\beta}(\cos\theta - \cos\beta)\,d\theta = r\int_{-\beta}^{\alpha}[y_1 + \frac{y_0 - y_1}{\sin\beta + \sin\alpha}(\sin\beta + \sin\theta)]\,d\theta$$

$$+ r\int_\alpha^\beta[y_2 + \frac{y_0 - y_2}{\sin\beta - \sin\alpha}(\sin\beta - \sin\theta)]\,d\theta.$$

Performing the indicated integration, and clearing of fractions, we obtain

$$y_0(2\beta\sin^2\beta - 2\cos\alpha\sin\beta + 2\sin\beta\cos\beta - 2\alpha\sin\alpha\sin\beta) + y_1(-\beta\sin^2\alpha$$
$$- \alpha\sin^2\alpha + \cos\alpha\sin\beta - \sin\beta\cos\beta + \alpha\sin\alpha\sin\beta + \beta\sin\alpha\sin\beta$$
$$- \sin\alpha\cos\alpha + \sin\alpha\cos\beta) + y_2(-\beta\sin^2\alpha + \alpha\sin^2\alpha + \cos\alpha\sin\beta$$
$$- \sin\beta\cos\beta + \alpha\sin\alpha\sin\beta - \beta\sin\alpha\sin\beta + \sin\alpha\cos\alpha - \sin\alpha\cos\beta)$$
$$= 2r(\sin^2\beta - \sin^2\alpha)(\sin\beta - \beta\cos\beta).$$

* Compare § 84.

96. **Third Equation of Condition.** — The third condition, in the modified from of § 59, is Σ D E . D B $=$ Σ D F . D B. Since D B $= r\,(\sin\beta - \sin\theta)$, this condition becomes, by multiplying the previous condition by D B,

$$r^3 \int_{\beta-}^{+\beta} (\sin\beta\cos\theta - \sin\theta\cos\theta - \sin\beta\cos\beta + \cos\beta\sin\theta)\, d\theta$$

$$= r^3 \int_{-\beta}^{\alpha} \left[y_1 (\sin\beta - \sin\theta) + \frac{y_0 - y_1}{\sin\beta + \sin\alpha} (\sin^2\beta - \sin^2\theta)\right] d\theta$$

$$+ r^3 \int_{\alpha}^{\beta} \left[y_2 (\sin\beta - \sin\theta) + \frac{y_0 - y_2}{\sin\beta - \sin\alpha} (\sin^2\beta - 2\sin\beta\sin\theta + \sin^2\theta)\right] d\theta,^*$$

which, when integrated and cleared of fractions, gives

$$\begin{aligned}
&y_0 (2\beta\sin^3\beta - \alpha\sin\beta - \sin\alpha\cos\alpha\sin\beta + 2\sin^2\beta\cos\beta - 2\alpha\sin\alpha\sin^2\beta \\
&\quad + \beta\sin\alpha + \sin\alpha\sin\beta\cos\beta - 2\cos\alpha\sin^2\beta) + y_1 (-\tfrac{3}{2}\sin^2\beta\cos\beta \\
&\quad + \cos\alpha\sin^2\beta - \beta\sin^2\alpha\sin\beta - \alpha\sin^2\alpha\sin\beta + \sin^2\alpha\cos\beta - \tfrac{1}{2}\sin^2\alpha\cos\alpha \\
&\quad + \tfrac{1}{2}\alpha\sin\beta - \tfrac{1}{2}\sin\alpha\cos\alpha\sin\beta + \tfrac{1}{2}\beta\sin\beta + \beta\sin\alpha\sin^2\beta + \alpha\sin\alpha \\
&\quad - \tfrac{1}{2}\alpha\sin\alpha - \tfrac{1}{2}\beta\sin\alpha + \tfrac{1}{2}\sin\alpha\sin\beta\cos\beta) + y_2 (-\tfrac{1}{2}\sin^2\beta\cos\beta \\
&\quad + \cos\alpha\sin^2\beta - \beta\sin^3\alpha\sin\beta + \alpha\sin^2\alpha\sin\beta - \sin^2\alpha\cos\beta + \tfrac{1}{2}\sin^2\alpha\cos\alpha \\
&\quad + \tfrac{1}{2}\alpha\sin\beta + \tfrac{3}{2}\sin\alpha\cos\alpha\sin\beta - \tfrac{1}{2}\beta\sin\beta - \beta\sin\alpha\sin^2\beta + \alpha\sin\alpha\sin^2\beta \\
&\quad + \tfrac{1}{2}\alpha\sin\alpha - \tfrac{1}{2}\beta\sin\alpha - \tfrac{3}{2}\sin\alpha\sin\beta\cos\beta) = 2r\sin\beta(\sin^2\beta - \sin^2\alpha) \\
&\quad (\sin\beta - \beta\cos\beta).
\end{aligned}$$

97. **Reduction of Equations.** — If the second equation of condition is multiplied by $\cos\beta$, and added to the first, there results an equation in which, as soon as we write $1 - \sin^2\alpha$ for $\cos^2\alpha$, and $1 - \sin^2\beta$ for $\cos^2\beta$, there will be found a common factor $(\sin^2\beta - \sin^2\alpha)$. This being cancelled out, there results (1.), § 92. The second equation again may be divided by 2, and then factored, by simple inspection, into (2.), § 92. Finally, the second equation of condition may be multiplied by $\sin\beta$, and subtracted from the third, when, upon factoring, we obtain (3.), § 92.

It will be seen that the solution of (4.), (5.), and (6.), § 92, for any given arch, and for several values of α, will not involve much work, owing to the recurrence of the known factors denoted by a, b, c, d, and e. As the arch may subtend any angle, it will not be expedient to go into calculations here for any special values of β. One case will be taken up later.

98. **Values of H, &c.** — When the desired ordinates for any arch are computed, we have the option of obtaining the values

---

* $\int \sin^2\theta\, d\theta = \frac{1}{2}(\theta - \sin\theta\cos\theta)$. See also note to § 84.

of H, of the vertical components of the abutment reactions, and of the ordinates for bending moment, either by graphical construction, or by formulæ similar to those applied to the parabolic rib. By noticing the expressions to be substituted for b, c, and k in the case of the circular arch with hinged ends, one can readily adapt the formulæ of § 63 and § 65 to the computations for this case. The ordinates to the circular arch will be the same as in § 85.

99. **Table of y_0, y_1, and y_2 for Semicircle.** — We may, however, obtain the ordinates y_0, &c., for a semicircle with comparative ease, and, as such a rib is sometimes used for large roofs, these values may be convenient. Semicircular masonry arches, having backing above the abutments, present a different case.

If α is taken as 20° or .3491, $\sin \alpha = .3420$, $\cos \alpha = .9397$, and $\frac{1}{2}\pi = 1.5708$; hence, in § 93, $g = .5117$, and (5.) and (6.) become

$$\tfrac{1}{2}(y_1 + y_2) = \frac{-.0792}{-.3646} r = .2172 r;$$

$$\tfrac{1}{2}(y_1 - y_2) = \frac{-.2977 \times .1333}{-.3646} r = .1088 r;$$

whence $y_1 = .326\, r$, and $y_2 = .108\, r$. By substitution in (1.), § 93, $y_0 = (1.5708 - .2187 - .0357)\, r = 1.316\, r$.

If similar computations are carried out for other values of α, we shall complete the following table for a semicircular rib with fixed ends:

α	y_1.	y_0.	y_2.
0°	.241 r	1.330 r	.241 r
10	.288	1.326	.183
20	.326	1.316	.108
30	.360	1.298	.011
40	.387	1.275	— .125
50	.413	1.245	— .330
60	.434	1.210	— .665
70	.455	1.170	— 1.333
80	.475	1.125	— 3.319

Other intermediate values can be obtained, if desired, by the formula for interpolation, § 45. The number of decimals it is desirable to use in any particular case will depend upon the value of r. The equilibrium polygons for these ordinates have been drawn in Fig. 25, and from them we get the different values of H, for a weight W at the several divisions, as shown in the accompanying stress diagram.

100. **Example.**—As an application of these results, let us draw the equilibrium curve for a semicircular arch of uniform section carrying only its own weight. As this weight is symmetrically disposed, $y_1' = y_2'$. By drawing the stress diagram of Fig. 25 to a sufficiently large scale, we shall find by measurement, that H, for a weight at the crown, 10°, 20°, &c., from the crown, will be .46, .44, .39, .31, .23, .14, .07, .02, and .01 W respectively. If we double all of these values except the one for a weight at the crown, and take the sum of the whole, we shall obtain for the horizontal thrust, $H' = 3.68\,W$ for 17 loads, each equal to W, at the 17 points of division in the whole arch.

To find y_1', multiply each y_1 by its H, remembering, that, when the weights are on the left of the crown, the values of y_2 in the table of § 99 become y_1, and that we may, therefore, before multiplying by H, add together y_1 and y_2 for each point except the crown, and then divide the sum of these products by H′, just obtained. (Compare § 67.) For example, for a load W on each of the two points distant 30° from the crown, $H\,y_1 + H\,y_2 = .31\,W\,(.360 + .011)\,r = .115\,r\,W$, the value of M at the abutments. Performing the operations, and taking the algebraic sum of the products, we get $.6225\,r\,W$ for the total moment at either abutment, and $\frac{.6225\,r\,W}{3.68\,W} = 0.17\,r = y_1' = y_2'$.

To construct the equilibrium curve, we divide the semicircle A C B, Fig. 26, into eighteen equal parts, each subtending 10°, and draw verticals through the points of division. Assume the weight of the arch to be represented by a vertical line of any convenient length. Since the loads are supposed to be concentrated at the points of division, one-eighteenth of the gross weight of the arch will be found at each of these points, and one-thirty-sixth at A and B; for A and B will each carry directly one-half of the adjacent division. Therefore, beginning and closing with one-thirty-sixth, space off the load-line into eighteenths; from the middle of the load-line lay off $H' = 3.68\ W = 3–0$, where $W =$ weight of one division, or $H' = \frac{3.68}{18} = .204$ of the whole weight of the rib. One-half of this load-line is 1–3. Lay off y_1' and $y_2' = .17\ r$, at A and B, and draw the sides of the equilibrium polygon parallel to the lines which radiate from the extremity of H′ to the points of division of the load-line, thus obtaining the curve E G D. The second half of the curve was obtained by spacing off 0′–3 to the left.

101. **Practical Application.** — Having at hand a wooden model of an arch-ring, representing the voussoirs, or stones, of a semicircular arch, we tried some experiments as tests of the accuracy of this method of analysis and of the correctness of these results. The arch is represented by Fig. 26, and consisted of forty-two independent voussoirs. The span, A B, of the middle line of the ring, 18 inches, was 13.09 times the thickness of the ring, and the structure would apparently just stand alone when left to itself: a slight additional weight at the crown would cause that part to sink, the haunches to move outwards, and the ring to fall in pieces. Considering that this arch, so long as it rested squarely on the faces at A and B, was fixed in direction, or not free to turn at the ends, we laid off at A E and B D the value of y_1 obtained in the last section, and drew the equilibrium polygon, as just described, on the centre line of the ring, beginning at D with a line parallel to 0–4. It will be noted that no line is used from 0 to 1; for the weight represented by 1–4 is directly supported at B; while the amount 4–5 is the weight concentrated on the first vertical just above D.

As the arch is a continuous ring, the weights may properly be concentrated at a greater number of points; so that finally the true equilibrium curve will pass through the vertices of the polygon we have just constructed: the difference between the two is unimportant, however, and is only appreciable near the crown. The bending moment at any point has been proved to be equal to H multiplied by the vertical ordinate between the centre line and the equilibrium curve, or, by § 10, also equal to T, the thrust along the tangent to the equilibrium curve, multiplied by the perpendicular from a point on the centre line to this tangent: therefore if we draw E F as this tangent, the bending moment at A will equal either H. E A, or the thrust along E F multiplied by the perpendicular from A. The direction of the thrust E F, if prolonged, cuts the springing joint very close to the outside edge: it will also be noticed that the equilibrium curve approaches quite near to the edge of the voussoirs at the crown G. Now, as we reminded the reader in § 11 that the force T, or 0′–1, at the distance F A from the centre line of the rib, is equal to the same force at the centre line and the couple which produces bending moment, conversely, the resultant of the pressure of this rib at the end A must cut the base in the prolongation of the line E F: in short, the tangent to the equilibrium curve at each point gives the direction and point of application of the resultant thrust at that right section of the rib to which it belongs, as ascertained by erecting a vertical from the middle point of the section.

102. **Limiting Position of Equilibrium Curve.** — If, as is usually the case, the intensity of the resisting force of the abutment at A is assumed to vary uniformly from one edge to the other, then, in case the resistance is zero at the inside edge and a maximum at the outside edge, the intensity at all points can be represented, as shown in the small sketch marked A′, by the ordinates of a triangle whose base is the breadth of a voussoir, and whose longest ordinate is the intensity of the pressure at the edge near F. The total pressure will be equal to the area of the triangle, and the resultant will pass through the centre of gravity of the triangle, cutting the base at one-third of its length from the outer edge. If there existed any tension near the inner edge, we should have two triangles, as shown in the

other sketch, the inclined line cutting the base at the point where the stress changed from tension to compression; and the resultant of the two stresses must, since they are of opposite kinds, lie outside of their separate resultants, and on the side of the greater one. This fact as to the position of the resultant of two opposite parallel forces was indicated in § 11, Fig. 2, and is one of the well-known properties of the lever, as proved in Mechanics.

Since, then, the resultant force, or the thrust on a section of the rib of Fig. 26, at A, B, and C, passes near the edge of the section, or, as it is often stated, outside of the *middle third* of the cross-section, we should expect to find tension at the inside edge of the joint at these points. As this model consists simply of wooden blocks placed in juxtaposition, a voussoir t exert tension on its neighbor at any point of contact, movement must immediately take place when the weight of the rib is allowed to act freely, rotation being set up about the outside edges at F, G, and Q. The crown will sink, the haunches will move outwards, and the arch may be expected to fall. The reader will remember that it was explained, in § 12, that an arch tends to move away from the equilibrium curve.

Since any material is compressible, it is probable that the assumption of a uniform variation of intensity of stress at any section will not be strictly true; that the stress may not be exerted over the entire surface of the *originally* plane joint; and that therefore the equilibrium curve may pass somewhat outside of the middle third of the joint without causing the arch to fall, although the joint will then open slightly at the edge where no pressure is exerted, by reason of the compression causing the joint to be no longer plane. But such an assumption gives an additional element of safety to a design, when the engineer so proportions his rib of rectangular section that the equilibrium curve of the load at any time shall never leave the limits of the middle third, and the tensile strength of the cement will not then be relied upon to assure stability.

103. Model as hinged at Three Points. — The arch of Fig. 26 stood when the string which at first passed around the exterior was removed, although a slight change of shape was observable. A close inspection, however, showed that the voussoirs at the crown and the two springings were then in contact only at the outer edges. The rotation at these joints, indicated in the last section as probable, had commenced; but, as soon as the rib became thus hinged at three points, it was in equilibrium. It is desirable, then, as a further test, to draw the equilibrium curve for this rib hinged at the crown and springings. As the change of shape and curvature was very little, the supposition that the weight of the voussoirs is concentrated along the arc K Q will be sufficiently near the truth for our purpose.

The half-weight being represented by 1–3, the first step is to find the value of H for this case, when the load is concentrated at intervals of ten degrees along the outer semicircle. We can avail ourselves of the formula of § 23, finding the different values of b by measurement, or from tables of sines, since $b = r \sin \theta$, and summing up the several amounts of H for the whole semicircle; or, as is done in this figure, we may use the principle explained in § 30, that any two sides of the funicular polygon, or two tangents to the equilibrium curve, will meet, when prolonged, on the vertical through the centre of gravity of the included weight. Since the arch is symmetrically loaded, the thrust at the crown will be horizontal, and therefore lie in the line K L; the centre of gravity of the quadrant arc K Q will be on the vertical line P L, drawn at such a distance, K L, from the crown as to satisfy the value for the ordinate from the centre of a circle to the centre of gravity of a circular arc, viz., $\frac{\text{radius} \times \text{chord}}{\text{length of arc}}$; and therefore the thrust at the springing will lie in the line Q L, drawn from Q to the intersection of the other two forces. As 1–3 represents the weight of one-half the arch, and the thrust at the crown is parallel to 3–0, a line from 1, parallel to Q L, will complete the triangle of forces, and, cutting the horizontal line at 9, will determine 3–9 to be the desired value of H. The equilibrium polygon can now be drawn from Q to K, its sides being successively parallel to lines radiating from 9, the first line being 9–4 and the last one 9–6. These lines are not drawn in the stress diagram. The other half of the polygon may be added, if desired.

It will now be seen, that, excepting the hinged points, the nearest approach of the equilibrium curve to the edge of a voussoir is at P, where it is still well within the rib, and consequently no further movement of the rib is to be expected. Another model, somewhat thinner than the one here illustrated, was experimented with, and would not stand. If the arch of Fig. 26 is slightly weighted at K, the joint at P begins to open on the outside, confirming the result, that the equilibrium curve here passes nearest to the inner edge. If it be objected that change of outline previously referred to carries the portion of the rib near P farther from the centre, so that the equilibrium curve may run nearer the edge than we have plotted it, we rejoin, that such a movement, carrying the centre of gravity, and hence the line P L, in the same direction, will cause Q L to make a slightly less angle with the vertical, diminishing the value of H, and moving the equilibrium curve also a little away from P.

104. **Model as hinged at Abutments.** — For the purpose of making an additional test of our results, we finally placed a small wire at A and B, thus hinging the rib on its centre line at these points. The equilibrium curve for one-half of the arch is A N K. The amount of H is determined by computation from the formula of § 85, which becomes, for a semicircular rib, $H = \frac{\cos^2 \alpha}{\pi} W$; and the summation for the whole arch, carrying W at intervals of ten degrees along the centre line, is $H = 2.86\,W$, laid off at 3–8. Radiating lines between 8–4 and 8–6 will enable one to draw A N K. The arch, when released, fell in ruins, and the first joint to open, on the outside at the haunch, was near N, lower than P in the former case.

We have dwelt on these curves at some length, as they give so good a confirmation of previous deductions and results, and as they will aid the reader in assuring himself that he understands the method of treatment. Such diagrams must, for accuracy, be drawn to quite a large scale, and the results will then be very satisfactory.

105. **Effect of Change of Temperature.** — It remains to find the effect of change of temperature on the circular rib with fixed ends. As was previously indicated in § 76, we must find the height $A\,G = B\,I = y_1$, at which the equilibrium line shall be drawn in Fig. 27, by the condition that the change of inclination at the abutments, or $\Sigma\,E\,F = 0$. If the notation of the angles subtended by portions of the arch is as before, and as marked in the figure, we have $E\,F = D\,E - y_1$, and

$$\Sigma\,E\,F = \int_{-\beta}^{+\beta} r\,(r\cos\theta - r\cos\beta - y_1)\,d\,\theta = 2r\,(r\sin\beta - r\,\beta\cos\beta - y_1\,\beta) = 0,$$

or

$$y_1 = r\left(\frac{\sin\beta}{\beta} - \cos\beta\right),$$

which becomes, for a semicircle,

$$y_1 = \frac{2\,r}{\pi} = 0.632\,r.$$

The first term of (1.), § 76, therefore becomes $\Sigma\,D\,E^2 - y_1 . \Sigma\,D\,E$. From § 84, $\Sigma\,D\,E^2 = r^3\,(\beta + 2\,\beta\cos^2\beta - 3\sin\beta\cos\beta)$, while $y_1 . \Sigma\,D\,E$ gives, as above, $r^3\left(\frac{\sin\beta}{\beta} - \cos\beta\right)(2\sin\beta - 2\,\beta\cos\beta)$; so that the first term reduces to $r^3\left(\beta + \sin\beta\cos\beta - \frac{2\sin^2\beta}{\beta}\right)$, and (1.), § 76, takes the form of

$$H_t . r^3\left(\beta + \sin\beta\cos\beta - \frac{2\sin^2\beta}{\beta}\right) = \pm\,2\,E\,I\,t\,e\,r\sin\beta.$$

$$H_t = \pm\,\frac{2\,E\,I\,t\,e}{r^2\left(\frac{\beta}{\sin\beta} + \cos\beta - 2\,\frac{\sin\beta}{\beta}\right)}.$$

For a semicircle, the formula for horizontal thrust simplifies into

$$H_t = \pm \frac{2\,E\,I\,t\,e}{r^2\left(\frac{\pi}{2} - 2\frac{2}{\pi}\right)} = \pm\, 6.45 \frac{E\,I\,t\,e}{r^2}.$$

The bending moments at the crown and springing can now be readily written, and compared with the values of § 90. The horizontal thrust for the semicircular rib fixed at the ends is five times as great as when the ends are hinged. The remarks of § 91 in regard to shear will apply equally well here.

For the Elliptic Rib, see § 153.

106. **Maximum Stress determined by Length of Ordinate; Rib of Rectangular Section.** — It may sometimes be convenient to have the means of determining from a simple inspection of a diagram, by noting the position of the equilibrium polygon, how much the maximum intensity of stress at any section exceeds the mean intensity. As the mean intensity $f = T \div S$ where T is the direct thrust and S is the area of cross-section, and is obtained at any point from the known value of the thrust in the side of the equilibrium polygon, the maximum intensity of stress will be readily found by multiplying by the proper ratio. The stress arising from bending moment in a solid section is always taken as uniformly varying (see Fig. 2). The combination of direct stress with that from bending moment will also give a uniformly varying stress.

Considering, first, the rib of rectangular cross-section, Fig. 28, we see, that if we call the intensity, A C, of direct stress unity, a bending moment which will produce a compression, D E, of unity at the upper extreme fibre, and a tension, C A, of unity at the lower extreme fibre, will bring the resultant stress at all points to the amounts indicated in the left-hand sketch, twice the mean intensity at one edge, and zero at the other. If the cross-section is treated by the method of Part I., "Roofs," p. 57, Fig. 24, in order to make an equivalent area of uniform stress equal to the maximum, we get the shaded area of the section on the left, which is evidently one-half of the whole section. The centre of gravity of this area, lying at one-third the height from the upper edge, will be the point of application of the resultant force on the cross-section. If the bending moment is reversed, the sketch will be inverted: hence, when the line of thrust, or the side of the equilibrium polygon, passes at *one-sixth* of the depth above or below the *axis* of the rib, the intensity of stress at that edge of the rib which is nearer the line of thrust will be twice the mean intensity.

If, again, the maximum intensity is to be thrice the mean, the line F G, starting at a distance B F = 3 B D, still cuts C D at its middle point in order to make the total tension from bending moment equal to the total compression from the same cause. Noting where F G cuts A B, we have the point of no stress at $\frac{3}{4}h$ from the upper edge of the section: hen
shaded areas are drawn as given in the section on the
the upper one for compression, the lower one for tension.
area of the upper one is $\frac{1}{2}b \cdot \frac{3}{4}h = \frac{3}{8}bh$: the lower one, being similar, but of one-third the altitude, has one-ninth the area of the other, or $\frac{1}{24}bh$. The difference is $\frac{1}{3}bh$, or one-third the area of the cross-section, as required if the maximum intensity is to be three times the mean. Letting these areas represent the forces, and taking moments about the upper edge, each force being applied at the centre of gravity of its triangle, we have for the position of the resultant, measured from the upper edge,

$$\frac{\frac{3}{8}bh \cdot \frac{1}{4}h - \frac{1}{24}bh \cdot \frac{11}{12}h}{\frac{1}{3}bh} = \frac{1}{6}h.$$

If, therefore, the line of thrust passes at $\frac{1}{6}h$ from the edge, or one-third the depth from the axis, the intensity of compression on the outside fibre nearer the line will be three times the mean compression, and at the other edge there will be a tension equal in magnitude to the mean stress.

In the same way it may be shown, that, when the line of thrust cuts the edge, the compression there will be B I, four times the mean, and the tension at the other edge will be A K, twice the magnitude of the mean stress. Thus it will be seen, that, for every one-sixth h that the line of thrust is distant from the axis, the compression on the square inch will be increased by unity on the side to which the line deviates, and diminished by unity on the other side, the mean compression being denoted by unity. This is indicated by the numerals marked on the sketches of Fig. 29.

107. **Rib of Two Flanges.** — If the rib is composed of two flanges and an open-work web, the stress in either flange is easily determined. If the line of thrust is in the axis, each flange will carry one-half of the direct stress. If the line of thrust passes through one flange, Fig. 30, that flange may be considered to carry all of the compression uniformly distributed, and the other flange to be under no stress; for the depth of the flange is so small, compared with the whole depth of the rib, that no error of importance is involved in considering the stress as uniformly distributed over the section of one flange. If the line of thrust passes without the rib a distance equal to its depth, we get, by taking moments at A, Fig. 30,

Thrust at C $\times$ 2 A B = Compression at B $\times$ A B;
or, *Compression* at B = 2 $\times$ direct stress.

If moments are taken at B, we find,

Tension at A = direct stress.

In the same way, if $B'C' = 2A'B'$,

Compression at $B' = 3 \times$ direct stress; Tension at $A' = 2 \times$ direct stress.

Hence we may draw a sketch for this rib similar to the one for the rectangular rib. The numerals here denote that one flange carries once, twice, &c., the *entire* direct stress. If the rib has a plate web, or is an I beam, the above method will give a good approximation to the true stresses. If the web is heavy, the method of the next section may be applied.

108. **Rib of Circular Section; General Construction.** — When the rib is of less simple section, we must return to the

graphical construction first referred to. As an instance, suppose the cross-section of the rib to be a circle. The variation of stress on a diameter, in the direction of deviation, is indicated by the left-hand sketch of Fig. 31, when the intensity of stress is twice the mean at one edge, and zero at the other. By constructing, according to the principles already laid down, Part I., "Roofs," the equivalent area of maximum intensity, we obtain the shaded area of the figure, and then we determine its centre of gravity by cutting out the area, and balancing it over a knife-edge. The deviation of the line of thrust from the centre of the circle, to make the maximum intensity twice the mean, and the minimum zero, is thus found, and proves to be one-fourth the radius.

By the construction of the other sketch, taking moments as in § 106, or reasoning by analogy, we find that the deviation, in order that the maximum shall be thrice the mean intensity of compression, and the tension at the other end of the diameter shall equal the mean stress, must be one-half the radius from the centre: hence, when the line of thrust cuts the edge, the maximum compression equals five times the mean, and the tension at the other extreme of the diameter is three times the mean compression. Thus we get the numerals and their positions, as given in the figure.

In a thin tube of circular, elliptical, or oval section, the maximum compression is nearly three times the mean intensity of direct stress where the equilibrium polygon cuts the surface of the tube; and a tensile stress equal in magnitude to the mean will then be found at the other end of the extremity of the diameter: hence proportionate distances of the side of the equilibrium polygon from the axis of the rib will give twice, four times, &c., the mean stress.

CHAPTER VIII.

ARCHED RIBS UNDER WIND PRESSURE: HORIZONTAL FORCES.

109. **Wind Pressure on an Inclined Surface.** — When arched ribs are used, as is often the case, for the support of a roof, the pressure of the wind, being normal to the surface, will have a different effect upon the arch from that caused by a simple weight or vertical force. While referring to Part I., "Roofs," p. 31, for some remarks about the action of wind on a roof, we will repeat here, that, if P equals the horizontal force of the wind on a square foot of a vertical plane, the perpendicular pressure on a square foot of a surface inclined at an angle i to the horizon may be expressed by the empirical formula, —

$$P \sin i^{1.84 \cos i - 1}.$$

If, then, the maximum force of the wind be taken as forty pounds per square foot, which is an amount sufficiently great for the purposes of a design, the perpendicular or normal pressure per square foot, on surfaces inclined at different angles to the horizon, will be: —

Angle of Roof.	Normal Pressure.	Angle of Roof.	Normal Pressure.
5°	5.2 lbs.	35°	30.1 lbs.
10	9.6	40	33.4
15	14.0	45	36.1
20	18.3	50	38.1
25	22.5	55	39.6
30	26.5	60	40.0

For steeper pitches, the pressure may be taken as forty pounds.

The resultant pressure at each of the joints in the rafter which is on the side of the wind is then ascertained as in the case of any roof. If the roof surface is curved, any short portion between two points where braces abut, or purlins rest, may be considered as straight, and the wind force will then be perpendicular to such portion; this pressure being the only force exerted by the wind. If the resultant pressure at each joint is then found, either graphically or otherwise, and is resolved into vertical and horizontal components, we may include the vertical component in the analysis already carried out in detail. The effect of the horizontal component remains to be considered.

110. **Form of the Equilibrium Polygon; Vertical ponent of Reaction.**—The tendency of such a force to di... the arch being resisted by the stiffness of the rib, the equilibrium polygon for a single horizontal force H, applied at any point I on the rib, Fig. 32, must, if the arch is hinged at the ends, be two straight lines, which start from the two springing points, and meet on the prolongation of the line of action of H; for the rib must be in equilibrium under H and the two forces at the abutments. In the case of the arch A C B of Fig. 32, the reactions at A and B must lie in the lines A G and B G, the point G being found on the horizontal line I G, but its location on that line being at present unknown. It will be evident, when we conceive H to be applied to the equilibrium polygon at G, that the side A G will be in *tension*, while G B is compressed: therefore the reaction at B will be a thrust, as usual, but that at A will be a tension; and, if H were the only applied force, the arch would tend to rise from the abutment A, and would require fastening down.

As H acts at a vertical distance I L above the springing line, the moment which tends to overturn the frame is H . I L. If we take either abutment as the axis of moments, the condition of equilibrium that the moments of exterior forces must balance

in which latter expression P, being constant, may be omitted. If b, as usual, denotes the horizontal distance of I, the point of application of the force, from the middle of the span, and c equals the half-span, we can find that

$$x_0 = \frac{b^3}{4c^4}(5c^2 - b^2) = \tfrac{1}{4}n^3(5 - n^2)\,c, \quad (2.)$$

when $b = nc$. We shall see that x_0, depending for its sign upon that of b, will always be laid off on the opposite side of the centre from b, since it is so first taken in the figure, and hence that H_1, the horizontal *tension*, is always greater than one-half of H. The value of x_0 is independent of k.

114. **Proof of Formula.**—Retaining the usual notation, we have $AL = c - b$, $LB = c + b$; and $GQ = IL = \frac{k}{c^2}(c^2 - b^2)$. If x denotes the horizontal distance, B D, to the abutment, from any ordinate, D E, on the right of I we have

$$DE = \frac{k}{c^2}(2cx - x^2), \text{ and } DF : DB = GQ : QB, \text{ or } DF = \frac{k}{c^2}(c^2 - b^2)\frac{x}{c - x_0}.$$

As $EK : EF = QB : GQ$, and $EF = DE - DF$, we have

$$EK = (DE - DF)\frac{QB}{GQ}, \text{ and } EK \,.\, DE = (DE^2 - DE \,.\, DF)\frac{QB}{GQ}.$$

Substituting the values of these quantities, we get

$$\Sigma EK \,.\, DE = \int \frac{k}{c^3}\left[(2cx - x^2)^2 - (2cx - x^2)\,x\,\frac{c^2 - b^2}{c - x_0}\right]\frac{c - x_0}{c^2 - b^2}\,dx$$

as the expression which is applicable from B to I. From A to I the abscissa E K will be limited by the line A G, which differs in inclination from B C. If x, however, is now reckoned from A to the right, and A Q, denoted by $c + x_0$, is used in place of Q B, we have an expression for the space from A to I. This expedient was used in previous sections. As A G is in tension while B C is compressed, these two portions of (1.), § 113, will have opposite signs, and, when integrated, must be equal: we may, therefore, in equating, strike out the common constant quantities, obtaining

$$(c - x_0)\int_0^{c+b}(4c^2x^2 - 4cx^3 + x^4)\,dx - (c^2 - b^2)\int_0^{c+b}(2cx^2 - x^3)\,dx$$
$$= (c + x_0)\int_0^{c-b}(4c^2x^2 - 4cx^3 + x^4)\,dx - (c^2 - b^2)\int_0^{c-b}(2cx^2 - x^3)\,dx.$$

Performing the indicated integration, we get

$$(c-x_0)\left[\tfrac{4}{3}c^2(c+b)^3-c(c+b)^4+\tfrac{1}{5}(c+b)^5\right]-(c^2-b^2)\left[\tfrac{2}{3}c(c+b)^3-\tfrac{1}{4}(c+b)^4\right]$$
$$=(c+x_0)\left[\tfrac{4}{3}c^2(c-b)^3-c(c-b)^4+\tfrac{1}{5}(c-b)^5\right]-(c^2-b^2)\left[\tfrac{2}{3}c(c-b)^3-\tfrac{1}{4}(c-b)^4\right],$$

which at once reduces to

$$\tfrac{16}{15}c^5 x_0=\tfrac{4}{3}c^3 b^3-\tfrac{4}{15}c\,b^5,$$

or

$$x_0=\frac{b^3}{4\,c^4}(5\,c^2-b^2).$$

115. **Another Proof.**—We may, if we please, find the desired distance x_0 by another method. Imagine the roof of Fig. 34 to have two equal but opposite forces, H, applied at the two points C and G in the same horizontal line. These forces, if acting alone, will tend to diminish the span of the roof; there will be no vertical forces; and as the bending moments caused by them, in case the rib did not rest upon abutments, would be directly proportional to E F, the change of span would be proportional to Σ E F . D E from C to G. When the rib is retained by abutments, one H will give rise to H_1 at A, and H_2 at B: the other H will cause H_2 at A, and H_1 at B. As H_1 is always opposite in sign to H_2, the resultant force at each abutment will be $H_1 - H_2$, and is manifestly a tension exerted by the abutment on the rib. The change of span due to $H_1 - H_2$ will be proportional to $\Sigma\,DE^2$ from A to B (compare § 74), and this change of span must offset the one from H.

If D is at a distance x from the middle of the span, and C is distant b from the same point, we have $DE = \frac{k}{c^2}(c^2 - x^2)$, and $EF = \frac{k}{c^2}(b^2 - x^2)$. Since the rib is acted upon symmetrically, we need only integrate from the middle to one side; and we therefore have, when we drop the common factor $\frac{k}{c^2}$,

$$(H_1 - H_2)\int_0^c (c^2 - x^2)^2\,dx = H\int_0^b (b^2 - x^2)(c^2 - x^2)\,dx,$$

or

$$(\mathrm{H}_1 - \mathrm{H}_2)\,\tfrac{8}{15}\,c^5 = \mathrm{H}\,(\tfrac{2}{3}\,b^3\,c^2 - \tfrac{2}{15}\,c^5). \quad (a.)$$

From the stress diagram of Fig. 33 we see that

$$\mathrm{H}_1 : \mathrm{H}_2 : \mathrm{H} = c + x_0 : c - x_0 : 2\,c;$$

whence

$$\mathrm{H}_1 - \mathrm{H}_2 = \mathrm{H}\,\frac{c + x_0 - c + x_0}{2\,c} = \mathrm{H}\,\frac{x_0}{c}.$$

Substituting this value in (*a.*) we get, as before, § 114,

$$x_0 = \frac{b^3}{4\,c^4}\,(5\,c^2 - b^2).$$

116. **Formulæ for H_1 and P.** — The value of H_1 is seen to be, from the above proportion,

$$\mathrm{H}_1 = \mathrm{H}\,\frac{c + x_0}{2\,c} = \mathrm{H}\left(\frac{1}{2} + \frac{x_0}{2\,c}\right) = \mathrm{H}\left[\frac{1}{2} + \frac{b^3}{8\,c^5}\,(5\,c^2 - b^2)\right].$$

We also have, from Fig. 33,

$$\mathrm{P} : \mathrm{H} = \mathrm{G\,Q} : \mathrm{A\,B} = \frac{k}{c^2}\,(c^2 - b^2) : 2\,c;$$

or

$$\mathrm{P} = \mathrm{H}\,\frac{k}{2\,c^3}\,(c^2 - b^2) = \mathrm{H}\,\frac{k}{2\,c}\,(1 - n^2).$$

The reader may now calculate, if desirable, numerical values of x_0, H_1, and P, for different values of b, as was previously done for vertical forces. The several values of x_0 for four different positions of H are plotted in Fig. 33.

117. **Shear and Direct Stress.** — The shear will undergo some modification when the force applied to the arch acts horizontally, instead of vertically. The stress diagram is, as we have seen, a triangle, whose base is H, and whose altitude is P, represented by 0 1 2 of Fig. 36. At A of the parabolic rib the *thrust* is 1–0: if 1–4 is drawn parallel to the tangent at A, and 0–8 perpendicular to it, 1–8 will be the direct thrust, and 8–0 the negative shear, on a right section at A. This shear will

diminish at successive sections until we reach a point where the tangent to the rib is parallel to A G, when the shear will be zero, and the direct thrust 1–0. Beyond this point the shear will be positive until we pass I. At the abutment B, there is a *tension* 2–0: if 2–7 is drawn parallel to the tangent at B, 2–9 will be the direct tension, and 9–0 the shear, again negative, on a right section at B. In the same way the shear just to the left of I will be 10–0, positive, and to the right of I, 11–0, negative. It will be remembered that positive shear acts upward on the left of any section.

118. **Shear Diagram.** — A shear diagram may be drawn for a rib under a horizontal force by a similar method to the one previously explained, showing the *vertical* shear which will be projected on each right section. Lay off at *a* the quantity P = 3–0 = *af*, which is the vertical component of the reaction at A, and as P is constant across the entire span, being, in fact, the only external vertical force, complete the rectangle *afdb*. The vertical component which is required at A to produce 1–4 is 3–4, laid off at *ae*; and at B is 3–7, laid off above the line at *bl*, because 0–2 is a tension. A load of uniform intensity horizontally being required to put a parabolic rib in equilibrium, and H_1 being constant as far as I, draw *ecg* through *c*, the middle point of *ab*, and draw *ln* so as to pass through *c*, if prolonged. Then will the vertical ordinates between the inclined lines and *fd* represent the shear on a *vertical* section, and the projection of these ordinates on the respective normal sections will be the shear in the web. Thus *ef* is 4–0, which gives by projection 8–0, *ig* is 0–5, and *in* is 0–6. As in previous diagrams, the ordinates will be measured from the inclined lines, positive above and negative below, as marked. The shear will change sign at the point of maximum bending moment, and it will plainly be equal to P at the crown of the arch.

If it is remembered that the abutment reaction at B is of the opposite kind to that at A, or to the usual reaction for a weight W, the rotation of the diagram on the right of *i*, from the customary position below the line to its present place above

$a\,b$, will be accounted for. The force H has been assumed on the right in Fig. 36, in order that this shear diagram may be compared with that of Fig. 8. The vertical shear from a normal force may be found from an addition of these two figures. Moment diagrams cannot be added together in the same way, as the values of H and H_1 or H_2 will not be the same in the two cases.

119. **Circular Rib hinged at Ends.** — The method of finding x_0, introduced in § 115, is easily applied to the circular rib hinged at the ends; while the process of § 114 is considerably more involved. Let the angle subtended, in Fig. 35, by the half arch of radius r be denoted by β; the angle from the crown to the point of application of the external horizontal force, H, be α; and the variable angle from the crown to any point be θ. Let H be applied at two opposite points at the same level, as shown by the arrows in the figure, and let the abutment reactions be $H_1 - H_2$. Then, by parallel reasoning to that of § 115, we have, if y denotes any ordinate, and a the ordinate to the point of application of H,

$$(H_1 - H_2)\int_0^\beta y^2\,ds = H\int_0^\alpha (y - a)\,y\,ds.$$

$$y = r(\cos\theta - \cos\beta);\quad a = r(\cos\alpha - \cos\beta);\ \therefore$$

$$(H_1 - H_2)\,r^3\int_0^\beta(\cos^2\theta - 2\cos\theta\cos\beta + \cos^2\beta)\,d\theta$$

$$= H\,r^3\int_0^\alpha(\cos^2\theta - \cos\theta\cos\beta - \cos\theta\cos\alpha + \cos\alpha\cos\beta)\,d\theta.$$

Performing the integration, we get

$$(H_1 - H_2)\left(\tfrac{1}{2}\beta - \tfrac{3}{2}\sin\beta\cos\beta + \beta\cos^2\beta\right)$$
$$= H\left(\tfrac{1}{2}\alpha - \tfrac{1}{2}\sin\alpha\cos\alpha - \sin\alpha\cos\beta + \alpha\cos\alpha\cos\beta\right).$$

As in § 115, $x_0 = \dfrac{H_1 - H_2}{H}\,c = \dfrac{H_1 - H_2}{H}\,r\sin\beta$: whence

$$x_0 = r\sin\beta\,\frac{\alpha - \sin\alpha\cos\alpha - 2\cos\beta\,(\sin\alpha - \alpha\cos\alpha)}{\beta - 3\sin\beta\cos\beta + 2\beta\cos^2\beta}. \quad (1.)$$

If the rib is a semicircle, $\beta = \frac{1}{2}\pi$; $\cos\beta = 0$; $\sin\beta = 1$; and (1.) becomes,

$$x_0 = \frac{2r}{\pi}(a - \sin a \cos a). \quad (2.)$$

120. **Formulæ for H_1 and P.** – The value of H_1 will be, as in § 116,

$$H_1 = H\frac{c + x_0}{2c} = H\left(\tfrac{1}{2} + \frac{x_0}{2r\sin\beta}\right)$$

$$= \tfrac{1}{2}H\left(1 + \frac{a - \sin a \cos a - 2\cos\beta(\sin a - a\cos a)}{\beta - 3\sin\beta\cos\beta + 2\beta\cos^2\beta}\right),$$

and

$$P = \frac{Ha}{2c} = \frac{\cos a - \cos\beta}{2\sin\beta}H;$$

or, for a complete semicircle,

$$H_1 = \frac{\frac{1}{2}\pi + a - \sin a \cos a}{\pi}H; \quad P = \tfrac{1}{2}\cos a\, H.$$

121. **Experimental Verification.** — The values of x_0, obtained above, can be readily shown to be true by turning the model previously referred to through an angle of ninety degrees. A moderately stiff wire carefully bent to a curve A G B, Fig. 37, symmetrical with regard to the point G (an arc of a circle being probably the easiest one to fashion), is suspended from points C and D by strings from A to C, and from B to D. If the string B D is doubled so as to pass on both sides of the wire above G, A G B will be prevented from swinging round. A thread from A to B will hinder the span from enlarging, and will indicate by its slackening when the span is narrowed. If, then, a weight is attached at E, and, the string at C remaining stationary, that at D is moved until B is vertically below A, as proved by plumbing the thread A B, C A, when prolonged, will be found to intersect B D at F in the vertical line E F, giving the desired value of x_0. The point of intersection will be slightly changed by the weight of the wire, as before suggested in § 81. It is worthy of note that, H now being an external pull on the rib, in place of the usual thrust, x_0 will, in Fig. 37, be found on the same side of the centre with H.

122. Parabolic Rib fixed at Ends; Formulæ for x_0, x_1, and x_2. — Referring to Fig. 38, we will suppose that the external force H is applied at I, on the left of this parabolic rib with fixed ends; that the desired equilibrium polygon is given by the lines L G and N G C; and that the abscissæ, at present unknown, are, A L $= x_1$, B N $= x_2$, and O Q $= x_0$, the latter being measured from the middle of the span, and all being considered as positive when laid off as shown in this figure. The rest of the notation agrees with that used before. It may be proved that the abscissæ have the following easily computed values:

$$x_1 = \frac{1}{3}\left(c + \frac{4\,b^2}{c-b}\right); \quad x_2 = \frac{1}{3}\left(c + \frac{4\,b^2}{c+b}\right); \quad x_0 = 2\,\frac{b^3}{c^2},$$

or

$$x_1 = \frac{1}{3}\,c\left(1 + \frac{4\,n^2}{1-n}\right); \quad x_2 = \frac{1}{3}\,c\left(1 + \frac{4\,n^2}{1+n}\right); \quad x_0 = 2\,n^3 c.$$

Several of these values, for different positions of H, are plotted in Fig. 38.

If b is given successive values from $0.1\,c$ to $0.9\,c$, these quantities will be found to be

b.	x_1.	x_0.	x_2.
$0.1\,c$	$0.35\,c$	$0.002\,c$	$0.35\,c$
.2	0.40	0.016	0.38
.3	0.50	0.054	0.43
.4	0.69	0.128	0.49
.5	1.00	0.250	0.56
.6	1.53	0.432	0.63
.7	2.51	0.688	0.72
.8	4.60	1.024	0.81
.9	11.17	1.442	0.90

If b exceeds $0.7\,c$, the point of intersection falls without the rib.

123. First Equation of Condition. — If we remark that Q G, Fig. 38, the ordinate to the line of action of H, will be equal to I S, or to $\frac{k}{c^2}(c^2 - b^2)$, and that R K $=$ D E, we may find the value of E K as follows:

$$EK = RN - DN;\ RN : RK = QN : QG, \text{ or } RN = \frac{RK \cdot QN}{QG};$$

therefore

$$EK = \frac{DE \cdot QN}{QG} - DN.$$

These quantities, in the notation employed, may be written, if x is measured from the right abutment,

$$DE = \frac{k}{c^2}(2cx - x^2);\ QN = c + x_2 - x_0;\ DN = x_2 + x;\ QG = \frac{k}{c^2}(c^2 - b^2).$$

As $\frac{k}{c^2}$ will be a common factor in the equations which follow, we shall omit it. Substituting these values, we shall get, as the expression to be summed from B to I, for the first condition,

$$\Sigma EK \cdot DE = \int_0^{c+b} \left[\frac{c + x_2 - x_0}{c^2 - b^2}(4c^2x^2 - 4cx^3 + x^4) - (x_2 + x)(2cx - x^2)\right] dx.$$

If x is measured from the left abutment, L Q substituted for Q N, and x_1 written for x_2, we get an expression which is applicable from A to I, or

$$\Sigma EK \cdot DE = \int_0^{c-b} \left[\frac{c + x_1 + x_0}{c^2 - b^2}(4c^2x^2 - 4cx^3 + x^4) - (x_1 + x)(2cx - x^2)\right] dx.$$

As in § 114, these two expressions will be equated to make the change of span zero, and upon performing the indicated integrations, and multiplying through by $c^2 - b^2$, we obtain

$$(c + x_2 - x_0)\left[\tfrac{4}{3}c^2(c+b)^3 - c(c+b)^4 + \tfrac{1}{5}(c+b)^5\right] - (c^2 - b^2)\left[c x_2 (c+b)^2 - \tfrac{1}{3}x_2(c+b)^3 + \tfrac{2}{3}c(c+b)^3 - \tfrac{1}{4}(c+b)^4\right] = (c + x_1 + x_0)\left[\tfrac{4}{3}c^2(c-b)^3 - c(c-b)^4 + \tfrac{1}{5}(c-b)^5\right] - (c^2 - b^2)\left[c x_1 (c-b)^2 - \tfrac{1}{3}x_1(c-b)^3 + \tfrac{2}{3}c(c-b)^3 - \tfrac{1}{4}(c-b)^4\right].$$

This equation, by reduction and factoring, may be written,

$$8c^5x_0 - (c^5 - 5c^3b^2 + 5c^2b^3 - b^5)x_1 + (c^5 - 5c^3b^2 - 5c^2b^3 + b^5)x_2 = 10c^3b^3 - 2cb^5. \quad (1.)$$

124. **Second and Third Equations of Condition.** — The second condition, that the change of inclination at the abutments shall equal zero, is $\Sigma EK = 0$, and the portion of this expression from B to I will be,

$$\Sigma EK = \int_0^{c+b} \left[\frac{c + x_2 - x_0}{c^2 - b^2}(2cx - x^2) - (x_2 + x)\right] dx,$$

from A to I we may write, as just explained,

$$\Sigma \mathrm{E K} = \int_0^{c-b} \left[\frac{c + x_1 + x_0}{c^2 - b^2}(2cx - x^2) - (x_1 + x)\right] dx.$$

Equating, integrating, and reducing, we get

$$(c + x_2 - x_0)\left[c(c+b)^2 - \tfrac{1}{3}(c+b)^3\right] - (c^2 - b^2)\left[x_2(c+b) + \tfrac{1}{2}(c+b)^2\right]$$
$$= (c + x_1 + x_0)\left[c(c-b)^2 - \tfrac{1}{3}(c-b)^3\right]$$
$$- (c^2 - b^2)\left[x_1(c-b) + \tfrac{1}{2}(c-b)^2\right];$$

$$4c^3 x_0 - (c^3 - 3cb^2 + 2b^3)x_1 + (c^3 - 3cb^2 - 2b^3)x_2 = 4cb^3. \quad (1.)$$

In writing the third condition, that the abutment deflection shall equal zero, or $\Sigma \mathrm{E K} \,.\, \mathrm{D B} = 0$, we must, if we use the values of E K already adopted, make D B equal to x on the right of I, and equal to $2c - x$ on the left of I. We then have, from B to I,

$$\int_0^{c+b}\left[\frac{c + x_2 - x_0}{c^2 - b^2}(2cx^2 - x^3) - (x_2 + x)x\right] dx,$$

and from A to I,

$$\int_0^{c-b}\left[\frac{c + x_1 + x_0}{c^2 - b^2}(4c^2 x - 4cx^2 + x^3) - (x_1 + x)(2c - x)\right] dx.$$

Equating these two expressions and integrating, we find that

$$(c + x_2 - x_0)\left[\tfrac{2}{3}c(c+b)^3 - \tfrac{1}{4}(c+b)^4\right] - (c^2 - b^2)\left[\tfrac{1}{2}x_2(c+b)^2 + \tfrac{1}{3}(c+b)^3\right]$$
$$= (c + x_1 + x_0)\left[2c^2(c-b)^2 - \tfrac{4}{3}c(c-b)^3 + \tfrac{1}{4}(c-b)^4\right]$$
$$- (c^2 - b^2)\left[2cx_1(c-b) + \tfrac{1}{2}(2c - x_1)(c-b)^2 - \tfrac{1}{3}(c-b)^3\right],$$

which reduces to

$$16c^4 x_0 - (7c^4 - 18c^2b^2 + 8cb^3 + 3b^4)x_1 + (c^4 - 6c^2b^2 - 8cb^3 - 3b^4)x_2$$
$$= -2c^5 - 4c^3b^2 + 16c^2b^3 + 6cb^4. \quad (2.)$$

From (1.), § 123, and (1.) and (2.) of the present section, we may readily eliminate x_0, obtaining

$$(c^3 - b^3)x_1 - (c^3 + b^3)x_2 = 2cb^3,$$

and

$$(c^2 - b^2)x_1 + (c^2 - b^2)x_2 = \tfrac{4}{3}c^3 + 2cb^2,$$

whence may be deduced the formulæ of § 122.

125. Formulæ for H_1 and P. — The values of H_1, H_2, and P, can now be scaled from the stress diagram, which will also give, if preferred, the proportion

$$H_1 : H_2 : H = c + x_1 + x_0 : c + x_2 - x_0 : 2c + x_1 + x_2,$$

or

$$H_1 = H\frac{c + x_1 + x_0}{2c + x_1 + x_2} = H\left[\tfrac{1}{2} + (5c^2 - 3b^2)\frac{b^3}{4c^5}\right] = \tfrac{1}{2}H[1 + \tfrac{1}{2}n^3(5 - 3n^2)].$$

H_1 will therefore always be greater than $\frac{1}{2}$ H.

Likewise we have, for the vertical component of the abutment reactions,

$$P : H = \frac{k}{c^3}(c^2 - b^2) : 2c + x_1 + x_2$$

or

$$P = H \cdot \tfrac{3}{8}k\,\frac{(c^2 - b^2)^2}{c^5} = \tfrac{3}{8}H\frac{k}{c}(1 - n^2)^2.$$

The shear diagram for this case will follow the explanation given in § 118.

126. Circular Arch fixed at Ends. — There remains to be considered the circular rib, fixed at the ends, under the action of an external horizontal force. The notation of the angles is the same as that previously used for the circular arch. As H is here applied at a point on the right side, x_0, measured from the middle of the span, will now lie on the left of the centre O. Then we will prove that

$$x_1 = \left[\frac{f}{a} - \frac{ab - de}{ac}\left(\frac{f}{a} + \sin\beta\right)\right]r; \quad (1.)$$

$$x_2 = \left[\frac{f}{a} + \frac{ab - de}{ac}\left(\frac{f}{a} + \sin\beta\right)\right]r; \quad (2.)$$

in which equations

$a = \cos\alpha - \cos\beta$, $\qquad d = \beta\sin\alpha - \alpha\sin\beta$,

$b = \alpha\beta - \sin\alpha\sin\beta$, $\qquad e = 1 - \cos\alpha\cos\beta$,

$c = \beta^2 - 2\sin^2\beta + \beta\sin\beta\cos\beta$, $\qquad f = \beta - \cos\alpha\sin\beta$.

It will be noticed that c is constant for a given arch. The value of x_0 can then be obtained from the equation

$$2(\sin\beta - \beta\cos\beta)x_0 - [\sin\beta + \sin a - (\beta + a)\cos a]x_1 + [\sin\beta - \sin a - (\beta - a)\cos a]x_2 = 2r\sin\beta(\sin a - a\cos a). \quad (3.)$$

The distance x_1 and x_2 will, in every case, be laid off outwards from the abutments, and x_0 will be plotted away from the side where the force is applied. In these formulæ, x_1 is on the opposite side of the arch from the applied force, as is also H_1. In any case it is easy to distinguish between numerical values of x_1 and x_2, or H_1 and H_2, if we notice that the larger value belongs to the abutment which is nearer to the point of application of xternal force.

Several of the equilibrium polygons have been drawn in Fig. 39 for a horizontal force applied at different distances from the crown. The angle β of this rib is 60°; and the computed values of the abscissæ, for H at points distant 10° successively from one another, are

a.	x_1.	x_0.	x_2.
10°	.3704 r	.0186 r	.4212 r
20	.4755	.0762	.5860
30	.5892	.2547	1.0345
40	.7291	.5950	2.1559
50	.8749	1.1339	5.9953

127. **First Equation of Condition.** — The processes to be followed are akin to those already given: although the work is somewhat more tedious, it presents no difficulty. As in § 123, we shall find that, Fig. 39, $EK = RN - DN = \frac{DE \cdot QN}{QG} - DN$. In the usual notation

$$DE = r(\cos\theta - \cos\beta), \qquad QN = r\sin\beta + x_2 + x_0,$$
$$QG = r(\cos a - \cos\beta), \qquad DN = r\sin\beta + x_2 - r\sin\theta.$$

We therefore have

$$EK = \frac{r\sin\beta + x_2 + x_0}{\cos a - \cos\beta}(\cos\theta - \cos\beta) - (r\sin\beta + x_2 - r\sin\theta)$$

on the right of I. Upon the left of I, since E′K now equals D′L — RL, this expression will change in sign; and, since we measure from L, we must substitute x_1 in place of x_2, must subtract x_0 in place of adding it, and must change the sign of $r \sin \theta$: hence, on the left of I,

$$\mathrm{EK} = -\frac{r \sin \beta + x_1 - x_0}{\cos \alpha - \cos \beta} (\cos \theta - \cos \beta) + (r \sin \beta + x_1 + r \sin \theta).$$

The first condition, invariability of span, will now give,

$$\Sigma_{\alpha}^{\beta} \mathrm{EK} \,.\, \mathrm{DE} + \Sigma_{-\beta}^{\alpha} \mathrm{EK} \,.\, \mathrm{DE} = 0,$$

or, multiplying by $\cos \alpha - \cos \beta$,

$$\begin{aligned} r\int_{\alpha}^{\beta} [& (r \sin \beta + x_2 + x_0) (\cos^2 \theta - 2 \cos \theta \cos \beta + \cos^2 \beta) \\ & - (\cos \alpha - \cos \beta) (r \sin \beta + x_2 - r \sin \theta) (\cos \theta - \cos \beta)] \, d\theta \\ & + r \int_{-\beta}^{\alpha} [(r \sin \beta + x_1 - x_0) (- \cos^2 \theta + 2 \cos \theta \cos \beta - \cos^2 \beta) \\ & + (\cos \alpha - \cos \beta) (r \sin \beta + x_1 + r \sin \theta) (\cos \theta - \cos \beta)] \, d\theta = 0. \end{aligned}$$

The integration is similar to that already given for the circular rib in the earlier sections. There results, upon bringing together common factors,

$$\begin{aligned} (\beta - 3 \sin \beta \cos \beta + 2 \beta \cos^2 \beta) \, x_0 - (\tfrac{1}{2} \beta + \tfrac{1}{2} \alpha - \tfrac{1}{2} \sin \beta \cos \beta - \tfrac{1}{2} \sin \alpha \cos \alpha \\ - \sin \alpha \cos \beta - \cos \alpha \sin \beta + \beta \cos \alpha \cos \beta + \alpha \cos \alpha \cos \beta) \, x_1 \\ + (\tfrac{1}{2} \beta - \tfrac{1}{2} \alpha - \tfrac{1}{2} \sin \beta \cos \beta + \tfrac{1}{2} \sin \alpha \cos \alpha + \sin \alpha \cos \beta - \cos \alpha \sin \beta \\ + \beta \cos \alpha \cos \beta - \alpha \cos \alpha \cos \beta) x_2 = r \sin \beta \, (\alpha - \sin \alpha \cos \alpha - 2 \sin \alpha \cos \beta \\ + 2 \alpha \cos \alpha \cos \beta). \quad (1.) \end{aligned}$$

128. **Second and Third Equations of Condition.** — The second condition, that $\Sigma_{\alpha}^{\beta} \mathrm{EK} + \Sigma_{-\beta}^{\alpha} \mathrm{EK} = 0$, similarly gives,

$$\begin{aligned} \int_{\alpha}^{\beta} [(r \sin \beta + x_2 + x_0) (\cos \theta - \cos \beta) - (\cos \alpha - \cos \beta)(r \sin \beta + x_2 - r \sin \theta)] d\theta \\ + \int_{-\beta}^{\alpha} [(r \sin \beta + x_1 - x_0) (- \cos \theta + \cos \beta) \\ + (\cos \alpha - \cos \beta) (r \sin \beta + x_1 + r \sin \theta)] \, d\theta = 0. \end{aligned}$$

From this equation we obtain, by integrating and factoring,

$$\begin{aligned} (2 \sin \beta - 2 \beta \cos \beta) \, x_0 - (\sin \beta + \sin \alpha - \beta \cos \alpha - \alpha \cos \alpha) \, x_1 \\ + (\sin \beta - \sin \alpha - \beta \cos \alpha + \alpha \cos \alpha) \, x_2 = r \sin \beta \, (2 \sin \alpha - 2 \alpha \cos \alpha). \quad (1.) \end{aligned}$$

The third condition, that $\Sigma_{\alpha}^{\beta} \mathrm{EK} \, . \, \mathrm{DB} + \Sigma_{-\beta}^{\alpha} \mathrm{EK} \, . \, \mathrm{DB} = 0$, will give, when we introduce the value of $\mathrm{DB} = r(\sin\beta - \sin\theta)$,

$$r\int_{\alpha}^{\beta} [(r \sin\beta + x_2 + x_0)(\cos\theta - \cos\beta)(\sin\beta - \sin\theta)$$
$$- (\cos\alpha - \cos\beta)(r\sin\beta + x_2 - r\sin\theta)(\sin\beta - \sin\theta)]\, d\theta$$
$$+ r\int_{-\beta}^{\alpha} [(r\sin\beta + x_1 - x_0)(-\cos\theta + \cos\beta)(\sin\beta - \sin\theta)$$
$$+ (\cos\alpha - \cos\beta)(r\sin\beta + x_1 + r\sin\theta)(\sin\beta - \sin\theta)]\, d\theta = 0.$$

Operating upon this equation also, we find that

$$(2\sin^2\beta - 2\beta\sin\beta\cos\beta)\, x_0 - (\sin^2\beta + \sin\alpha\sin\beta - \tfrac{1}{2}\cos^2\beta - \tfrac{1}{2}\cos^2\alpha$$
$$+ \cos\alpha\cos\beta - \beta\cos\alpha\sin\beta - \alpha\cos\alpha\sin\beta)\, x_1 + (\sin^2\beta - \sin\alpha\sin\beta$$
$$+ \tfrac{1}{2}\cos^2\beta + \tfrac{1}{2}\cos^2\alpha - \cos\alpha\cos\beta - \beta\cos\alpha\sin\beta + \alpha\cos\alpha\sin\beta)\, x_2$$
$$= r\sin\beta\,(2\sin\alpha\sin\beta - \cos^2\alpha + \cos\alpha\cos\beta - 2\alpha\cos\alpha\sin\beta)$$
$$+ r\beta(\cos\alpha - \cos\beta). \quad (2.)$$

129. **Reduction.** — From (1.), § 127, and (1.) and (2.), § 128, we can determine the desired quantities x_0, x_1, and x_2, by any of the usual steps for elimination. If the second equation of condition is multiplied by $\sin\beta$, and then subtracted from the third, there will result

$$(\tfrac{1}{2}\cos^2\beta - \cos\alpha\cos\beta + \tfrac{1}{2}\cos^2\alpha)(x_1 + x_2)$$
$$= r\sin\beta(\cos\alpha\cos\beta - \cos^2\alpha) + r\beta(\cos\alpha - \cos\beta),$$

which, upon being divided by $\tfrac{1}{2}(\cos\alpha - \cos\beta)$, becomes

$$(\cos\alpha - \cos\beta)(x_1 + x_2) = 2r(\beta - \cos\alpha\sin\beta). \quad (a.)$$

Again: the second equation may be multiplied by $\cos\beta$, and added to the first, after which the values of x_0 from the new equation and from the second equation of condition may be equated. If we then clear of fractions, and factor the resulting equation, it may be written

$$[a(b - c) - de]\, x_2 + [a(b + c) - de]\, x_1 = -2r\sin\beta(ab - de), \quad (b.)$$

while equation (a.) will be

$$a(x_1 + x_2) = 2fr; \quad (c.)$$

in which equations the literal coefficients stand for the quantities already given in § 126.

From (*b*.) and (*c*.) it is easy for one to obtain the half sum and the half difference of the two unknown quantities, and thence equations (1.) and (2.), § 126. Equation (3.) is identical with (1), § 128.

130. **Formulæ for H_1, &c.; Semicircular Arch.** — To find the values of H_1, H_2, and P by formula, we make use of similar expressions to those of § 125. The figure gives us

$$H_1 : H_2 : H = r \sin\beta + x_1 - x_0 : r \sin\beta + x_2 + x_0 : 2 r \sin\beta + x_1 + x_2;$$

or

$$H_1 = H \frac{r \sin\beta + x_1 - x_0}{2 r \sin\beta + x_1 + x_2} = \frac{a}{2r} \cdot \frac{r \sin\beta + x_1 - x_0}{\beta - \sin\beta \cos\beta} H.$$

$$P : H = r(\cos a - \cos\beta) : 2 r \sin\beta + x_1 + x_2 = a r : 2 r \sin\beta + \frac{2fr}{a};$$

or

$$P = \tfrac{1}{2} \frac{a^2}{a \sin\beta + f} H = \tfrac{1}{2} H \frac{(\cos a - \cos\beta)^2}{\beta - \sin\beta\cos\beta}.$$

If the arch subtends a semicircle, $\beta = \frac{1}{2}\pi$, $\sin\beta = 1$, $\cos\beta = 0$, and the preceding values are much simplified. Without writing them in detail, it will be sufficient to indicate that then

$$a = \cos a, \qquad c = \tfrac{1}{4}\pi^2 - 2, \qquad e = 1,$$
$$b = \tfrac{1}{2}\pi a - \sin a, \qquad d = \tfrac{1}{2}\pi \sin a - a, \qquad f = \tfrac{1}{2}\pi - \cos a.$$

131. **Sign of Bending Moment.** — In determining the sign of the bending moment at any point when the arch is acted upon by a horizontal force, it will be well for the reader to recollect, that, when there is a thrust along any portion of the equilibrium polygon, the arched rib tends to move away from the polygon, but, when there is tension in any portion, the arch moves towards the polygon. This tendency to move in one direction or the other is easily fixed in the mind, if one thinks of the alteration of curvature of a bent wire when a force is applied at each end in the line joining the two ends. The same thing was noticed in the suspended arch of Fig. 1 and in those under vertical forces. Therefore, in Fig. 32 and the following

ribs, the arch tends to approach the tension side of the equilibrium polygon, and to recede from the compression side. If then, as before, that moment which makes any portion of the rib less curved, or which, if exerted on a beam supported at both ends, would make it concave on the upper side, be called positive, the areas of — M will occur between B and C in Figs. 32 and 33, and those of + M will be found between C and A. Ribs fixed at the ends will be strained similarly. In Fig. 38, for example, the area to the right of B will give + M; from the point where N G crosses the rib to C there will be — M, which then changes to + M on the left of C, and to — M, when the polygon crosses the rib above A.

132. **Example of Normal Forces.** — As we have now ascertained the values of the abutment reactions when a rib is acted ... a horizontal force, we will show, by an example, that ...is horizontal and vertical forces which are exerted at ...me at different points of the rib may be provided for in one polygon, without the necessity for separate treatment of the horizontal and vertical components into which the normal or oblique external forces can be decomposed. We will suppose that a parabolic rib of 100 feet span and 50 feet rise is to be used as a principal to carry a roof, and that it is desired to ascertain the bending moments arising from the action of the wind upon one side. We will take the case where the rib is fixed at the ends as being less simple. After this discussion, the reader will have no difficulty in applying a similar treatment to other ribs.

Let the rib be represented by A C B, Fig. 40, and let us suppose that the normal wind pressure is directly resisted by the flanges and bracing of the rib at points D, E, F, and G, at which purlins rest, and which are distant 40 feet, 30 feet, 20 feet, and 10 feet horizontally from the middle of the span. The amount of the pressure N_2 at E will be the total or resultant of the distributed pressure on *m n*, the points *m* and *n* being taken midway of the spaces on each side of E. There will be no error of consequence in assuming that the wind pressure on *m n* is perpendicular to the straight line *m n*, or to the tangent of the parabola at E.* To find this tangent, draw E E′ horizontally, make C E″ = C E′, and E E″ will be the desired tangent. The tangents at the other points are found in the same way. The angle E′ E E″ is very nearly 50°; the intensity of wind pressure, by the table of § 109, is 38 pounds on the square foot of roof; and if the principals are 10 feet apart, and *m n* is 15½ feet, the total normal force N_2 at this point will be $38 \times 10 \times 15\frac{1}{2} =$ 5,890 pounds. For the four points we therefore find in detail

						N.	V.	H.
1	58°	40 ×	19 ×	10 =		7,600 lbs.	4,000 lbs.	6,400 lbs.
2	50	38	15½	10		5,890	3,800	4,500
3	38½	32	13	10		4,160	3,200	2,600
4	22	20	11	10		2,200	2,000	900

These normal forces are plotted on the figure, and then decomposed graphically into their vertical and horizontal components, which, scaled to the nearest one hundred pounds, are found above in the columns headed V and H. The figure and diagrams are drawn to scales of forty feet and ten thousand pounds equal one inch.

133. **Finding the Reactions.** — The next step will be to find the values of H_1, H_2, P_1, and P_2, for the above forces. First, upon referring to § 64, we see that a vertical force at E, Fig. 40, 0.6 *c* from the middle of the span, will cause a vertical reaction of 0.896 V at A, one of 0.104 V at B, and will give rise to H, at each abutment, of the amount $0.192 \frac{c}{k} V = 0.192 V$. We also see, by the table of § 62, that the ordinate at A will be $-0.667 k$, and at B $+0.333 k$, for the same force at E; and we can then obtain the values of M at the abutments arising from V by multiplying these ordinates by $H = 0.192 V$, just ascertained. The computations for the four loaded points may be grouped together as follows:

* If preferred, analyze the wind pressures as in Part I., Roofs, p. 44.

	V.		P_1.		H.	
1	4,000 × 0.972	=	3,888 lbs.	V × .0607 =	243 lbs.	
2	3,800	0.896	3,405	.1920	730	V = 13,000 P_1' = 11,098 P_2' = 1,902 lbs.
3	3,200	0.784	2,509	.3308	1,059	
4	2,000	0.648	1,296	.4320	864	
	13,000		P_1' = 11,098 lbs.		H′ = 2,896 lbs.	

	H.	y_1.	M_1.	y_2.	M_2.
1	243 ×	−2.000 k =	−24,300 ft. lbs.	0.370 k	+ 4,495 ft. lbs.
2	730	−0.667	−24,333	0.333	12,167
3	1,059	−0.222	−11,767	0.286	15,144
4	864	0.000	000	0.222	9,600
Totals . . .			M_1' = −60,400 ft. lbs.		M_2' = + 41,406 ft. lbs.

is to be understood that y_1, P_1, and M_1 refer to the left abutment, the others, to the right abutment.

From § 122 and § 125 we now compute the reactions from the horizontal forces at the four loaded points, and the accompanying bending moments:

	H.		± P.			
1	6,400 ×	.0486 =	311 lbs.	H × 0.894 =	5,722 lbs.	
2	4,500	.1536	691	0.712	3,204	H = 14,400 H_1' = 10,872 H_2' = + 3,528 lbs.
3	2,600	.2646	688	0.572	1,487	
4	900	.3456	311	0.510	459	
	14,400					

Totals, P′ from H's = ± 2,001 lbs. H_1' = −10,872 lbs.

	P.	x_1.	M_1.	x_2.	
1	311 ×	4.600 c =	−71,530 ft. lbs.	0.807 c	+ 12,549 ft. lbs.
2	691	1.533	−52,976	0.633	21,870
3	688	0 689	−23,702	0.486	16,718
4	311	0.400	− 6,220	0.378	5,878
Totals			M_1' = −154,428 ft. lbs.		M_2' = + 57,015 ft. lbs.

The final abutment moments will be

$$M_1' = -60,400 - 154,428 = -214,828 \text{ ft. lbs.}$$

$$M_2' = 41,406 + 57,015 = +98,421 \text{ ft. lbs.}$$

The components of the reaction at A are, if thrusts are considered positive,

$$P_1' = P_1 - P = 11,098 - 2,001 = +9,097 \text{ lbs.}$$

$$H_1' = H + H_1 = 2,896 - 10,872 = -7,976 \text{ lbs.}$$

The components at B will be

$$P_2' = P_2 + P = 1,902 + 2,001 = +3,903 \text{ lbs.}$$

$$H_2' = H + H_2 = 2,896 + 3,528 = +6,424 \text{ lbs.}$$

The arrows at A and B show these reactions. If the rib consists of chords and bracing, the stresses on the pieces can be found by a diagram like Fig. 21, Part I., "Roofs," care being taken to have the stresses in the two flanges at the abutment give the proper reaction (see § 195). If the equilibrium polygon is to be drawn, from which to find bending moments and chord stresses, we need the point of beginning for the polygon.

The abscissa, or ordinate to the equilibrium polygon at A, will be found by dividing the total M at that point by P_1' or H_1'; and similarly for the abutment B; thus,

$$x_1' = \frac{-214,828}{+9,097} = -23.6 \text{ ft.} \qquad x_2' = \frac{+98,421}{+3,903} = +25.2 \text{ ft.}$$

$$y_1' = \frac{-214,828}{-7,976} = +27.0 \text{ ft.} \qquad y_2' = \frac{+98,421}{+6,424} = +15.3 \text{ ft.}$$

As in previous examples, the ordinate at one abutment alone is needed; but the others are useful as a check on the accuracy of the drawing.

134. **Equilibrium Polygon; Bending Moments.** — We may now proceed to draw the stress diagram. Lay off 1–2, 2–3, 3–4 and 4–5, parallel successively to the external forces at G, F, E,

and D, and equal to the calculated amounts by any desirable scale; make 5–6 = H_1', and 6–0 = P_1', so that 5–0 shall repre- ent the reaction at A in the proper direction as expressed by e signs obtained above, P_1' being a compression, and H_1' a ten- ı; finally, lay off 0–7 = P_2', and 7–1 = H_2', giving 0–1 for reaction at B. The closing of 0–1 on the point 1 proves t the diagram has been drawn with care. Having drawn } = + y_2', or B R = + x_2', draw through Q or R a line par- l to 0–1, as far as O, where it meets the normal force at G. n draw O L parallel to 0–2, to cut the force N_3 at L. Fol- with L K and K I, parallel to 0–3 and 0–4, closing with a ne through I, parallel to 0–5, which, if the polygon has been accurately drawn, will make A W = y_1', as recently computed, or A U = $-x_1$.

As neither H nor P is constant for *oblique* forces on an arch, the bending moment at any point will equal the product of the force acting along a side of the polygon just drawn multiplied by the perpendicular from the point to the side: thus the bending moment at E is E S × (0–3), or E T × (0–4). If the external forces had been considered as applied at a greater number of points, or as distributed along the principal rafter itself, we should have obtained a polygon which approached nearer to a regular curve, and such a curve has been sketched through the vertices of the polygon just drawn.

135. **Equilibrium Polygons for the Vertical and Horizontal Components.** — Since most of the needful data have already been obtained, we have thought it expedient to draw the equilibrium polygons for the vertical and horizontal components separately, so that they may be compared with each other and with the polygon for normal forces. If a horizontal and a vertical line are drawn from 1 and 5, the components H and V can be at once projected upon them. Upon laying off H_1, and plotting P, we shall locate the pole 0″; and 0″–2″, 0″–3″, &c., will be parallel to the lines of the polygon for horizontal forces. In the same way, P_1 and H for vertical forces will determine 0′. The value of y_2 will be found, upon dividing the M_2 which

comes from V by H, to be 14.3 feet, giving the starting-point just below Q. Upon drawing the polygon so that the angles are made at the verticals through the loaded points, we obtain the broken line which finally runs below A. This ordinate y_1 may be verified. If M_2 from the H's is divided by P, we have $x_2 = 28.5$ feet, an ordinate a little longer than B R. The polygon, if now drawn, will be the broken line which passes near E′, and extends to a considerable distance, 77.2 feet, to the left of A. All the sides of this polygon except the first are in tension.

136. **Shear and Direct Stress.**—To complete this example, the normal shear at the middle of each division is found, and at the same time the direct stress. The small letters *l*, *m*, *n*, &c., mark the middle of each division. Draw 0–*l* in the stress diagram, parallel to the tangent at *l* in the rib, and 5–*l* perpendicular to it; then will 5–*l* be the normal shear at *l*, and *l*–0 the direct thrust. To satisfy ourselves in regard to the sign of this shear, we note that 5–0 is the thrust in the side U I of the equilibrium polygon, and will therefore be the resultant force on the left of any section between A and D; the forces 5–*l* and *l*–0, in the directions named, will be its components, also on the left of the section *l*: hence we have *positive* shear and a direct thrust. In the same way at *m*, since 4–0 is the thrust in I K, 4–*m* will be the positive shear, and *m*–0 the direct thrust. Between *m* and *n* the shear changes sign; for at *n* we find 3–*n* and *n*–0, the former being drawn *down*, instead of *up*. Passing on, we see that the shear again changes between *r* and *s*, because 1–*r* and 1–*s* run in opposite directions. As noted before, this change of sign occurs at points of maximum bending moment.

137. **Vertical Shear Diagram.**—We may draw a vertical shear diagram, if desired, and from that obtain the normal components; but it is not so conveniently constructed in the case of several forces which are always applied together as for a case of a single load. If *a b* represents the span, P_1' or 6–0 is laid off at *a w*, upwards as usual; then the subtraction of V_1 at D, or 4′–5, brings us to the line *d*; thence a step is made to *e*, to *f*, and finally to *g*, closing at *b* with 0–7, the reaction at B. The horizontal line below *a b* cuts off P, or 0″–3″, so that the vertical components shown in the line 5–1′

might be considered as laid off from this lower line, and the constant quantity P, due to the horizontal components, then subtracted. As the thrust at B is 0–1, a line drawn through 0, parallel to the tangent at B, will cut off from a vertical line drawn from 1 as much vertical force as is required, in addition to 0–7, to give a resultant in the direction of the rib at B. The amount so determined is laid off at *q′ r′*. Since it has been shown that all inclined lines are drawn towards the middle of the span *c*, and are uninterrupted until an external force is encountered, we draw through *c* the line *r′ c s*

In a similar way, a line 0–10 from 0, parallel to the tangent at A, will cut the vertical through 5 at a distance 5-10, equal to *w u;* a line from 0, parallel to the tangent at D, will cut off the distance from a vertical through 4, which is plotted from *d* to *k;* one parallel to the tangent at E will cut off 3–8, which is plotted at *e o;* and the tangent at F gives 0–9, so that 2–9 is laid off at *f p*. If inclined lines are drawn through the points thus found, running towards the point *c*, the diagram will be completed. Normal components of the ordinates between the two sets of lines just constructed, measured above *l*, *m*, *n*, &c., will agree with the values of the last section, —positive when above the inclined lines, negative when below.

CHAPTER IX.

STONE ARCHES.

138. **Location of Equilibrium Curve determines Thickness of Voussoirs.**—Stone arches may be treated as belonging to the class of ribs with fixed ends, as the voussoirs have sufficient breadth at the skew-backs to make a firm bearing. We can, then, for a given rise, span, and distribution of steady and travelling load, draw the equilibrium curve, and thence determine the required thickness of the arch-ring. To repeat what was mentioned incidentally earlier: if no reliance is placed upon the tenacity of the cement, and if the intensity of pressure at a joint between any two voussoirs or arch-stones is considered to vary uniformly from the outside to the inside edge, the extreme case of deviation of the resultant pressure from the middle of the joint consistent with safety will occur when the pressure is zero at one edge. As the varying intensity of pressure will be represented by the ordinates to an inclined line which passes through the point where the pressure is zero, the total pressure will be equal to the area of a triangle, and the resultant will pass through the centre of gravity of the triangle, or at a distance of one-third the breadth of the ring from that edge where the pressure is most intense. Since the equilibrium curve is the locus of the resultant force at each joint, the condition that the pressure shall never be less than zero at any point, or that there shall be no tension, is equivalent to requiring that the equilibrium curve shall never pass beyond the middle third of the

arch-ring, however the distribution of the load may be varied: hence, when the equilibrium curves are drawn, the thickness of the voussoirs is readily determined. The tensile strength of the cement after it has become firm, and any deviation from the assumption that the force between two stones must be distributed over the whole joint, increase the safety of the structure, and thus give what is akin to the factor of safety in other cases.

139. **Intensity of Pressure.** — When the stability of the arch-ring is thus assured, it is an easy matter to find the greatest intensity of pressure, and hence to see whether the material proposed for the arch will have strength enough. When the equilibrium curve passes through the centre of the joint, the pressure on the square inch will be found by dividing the thrust at that joint by the area of the bearing surface. If the curve touches the extreme limit, the edge of the middle third, the most intense pressure, at the edge of the joint nearest to the curve, will be twice the mean pressure; for the height of the triangle whose ordinates represent the varying intensities is twice its mean ordinate. In some rare cases, where the span is large, and the stone is of a weak quality, we may have to increase the depth of the arch-ring in order to provide sufficient strength.

140. **Circular Arch; Load for Equilibrium.** — Although the curve of the arch-ring may be any one of a number of forms, the circular arch is the more common type, and we have therefore thought it best to take such an arch as an example of this method: the steps will apply to any form. The Gothic arch will be classed with the example of § 194. If the load is entirely, or almost entirely, steady, as in the aqueduct or canal bridge, it will be advisable, on the score of economy, to find that distribution of the load which shall cause the equilibrium curve to coincide with the centre line of the arch-ring. Then, by arranging the filling and the empty spaces above the arch-ring so as to conform to that distribution, the voussoirs can be made of moderate depth.

Thus, if B C, Fig. 45, be one-half of an arch which it is desired to load in this way, divide it, by vertical lines, into quite a large number of parts, equal horizontally. If the divisions are small, the areas of these portions between the soffit of the arch and the upper line may be considered trapezoids, and the middle ordinate of each division will be proportional to its volume for unity of thickness, and to its weight, if homogeneous. It is then evident, that, if there is to be no bending moment at any point, the equilibrium curve must coincide, either with the tangents to the centre line of the ring at these loaded points, or with the chords drawn between these points, according as the first loaded point is taken at half a division's distance from the abutment, or at the abutment itself. See Part II., "Bridges," § 58. Let this weight be concentrated in imagination, on each middle ordinate.

Upon drawing, from any point 0, radiating lines parallel the tangents, or perpendicular to the radii, at the successive points of division, and cutting them all by a vertical line 1–12 at any convenient distance, loads in each division, supposed to be concentrated at the intersection of the above tangents,* and proportional to the several portions of the vertical line intercepted by the inclined lines, will be the ones required for equilibrium; and the distributed loads spread over all of each division, or, in other words, a continuous load over the whole arch, will thus be found. If 1–2 is placed at such a distance from 0 that it will represent, by a convenient scale, the mean depth, as well as the weight of the load, in the first division on the right of C, 2–3, 3–4, &c., will represent the required depth of loading in the succeeding divisions. As the angle made by 0–2 with the horizontal line is the same as that subtended at the centre by the first division near C, there is no difficulty in finding, by calculation, the exact length of 0–1, when 1–2 is given, in case the angle at 0 is too acute to give any accurate result graphically. In our figure the depth of the load at the

* The tangents will not intersect exactly in the middle of each division.

crown was assumed to be five feet, and the intercepted portions of the vertical line were then plotted from the points where verticals at the middle of each division would cut the centre line of the arch. The curved line drawn through the upper ends of these ordinates will then show the desired amount of homogeneous load to be spread over the arch to produce equilibrium.

141. **Limiting Angle for Arch-Ring without Backing.**— It is now worthy of notice, that, while the required depth of loading increases but slowly for some distance after we leave the crown, when we reach the haunches, the ordinates rapidly lengthen, and the curve through their upper ends will finally become vertical, if the arch springs vertically from the abutment. This point was also referred to in § 89. It is apparent, therefore, that it is not practicable to so load with vertical forces a circular arch, beyond a certain distance from the crown, that the line of thrust shall coincide with the centre line of the arch-ring. As the roadway must not deviate greatly from a horizontal line, we see, that, for an arch extending 60° each way from the crown, the amount of material as heavy as masonry required over the springing will fill all of the available space, and, when the spandrel filling is lighter, the limiting angle will probably be in the neighborhood of 45°. In ordinary cases of loading, the equilibrium curve will deviate so much from the centre line in this portion of the rib as to require very deep voussoirs to retain the curve within the middle third when the attempt is made to extend the unassisted arch-ring much farther. It is customary, therefore, to carry the masonry backing, in horizontal courses, up to the neighborhood of the point where the arch-ring is inclined at an angle of 45°: below this point any attempt of the arch-ring to move outwards under the thrust of the upper portion is immediately resisted by the backing, and the arch will be designed as if the springing points were at the joints level with the top of this masonry backing. The portion below really forms a part of the abutment.

142. **Example; Data.**—In accordance with the above statements, and as an example of the application of preceding principles, we propose to design a circular segmental arch of stone, for a railroad bridge, which shall subtend 100°, with a radius, for the centre line of the voussoirs, of 100 feet, making the span, from centre to centre of skew-backs, about 153 feet, and the rise about 36 feet. The rolling load will be 3,000 pounds per running foot of track, and the width of the bridge over which this load is distributed will be ten feet. The backing will be carried up to the point where the rib is inclined at 45°, and the remainder of the spandrel will be filled with such material, or will have such an amount and distribution of empty spaces, that it shall weigh, on the average, one-half as much per cubic foot as does the masonry of the arch-ring. The equilibrium curve for steady load will now first be found; then such possible combinations of rolling load will be discussed as will increase the deviation of the steady load curve at those points where it already deviates most from the centre line of the arch-ring; and, finally, the necessary depth of the voussoirs will be determined by the rule suggested in § 138. The depth of the voussoirs at the crown is assumed, in our present ignorance of the final dimensions, at five feet; two feet of filling, earth or some other material, is added at that point, and the horizontal line drawn seven feet above the soffit at the crown will be the upper boundary of the spandrel filling. If, then, the arch-ring is taken at a uniform thickness of five feet, as shown at A C, on the left half of Fig. 45, the depth of a homogeneous load equal to stone will be found by shortening each ordinate above the arch ring one-half. Thus was obtained the curve D E. By dividing the area between this curve and the soffit into small portions by vertical lines, we may find the weight to be concentrated on the several assumed loaded points of the arch-ring.

143. **Calculations for Steady Load.**—From the equations of § 92, after making $\beta = 45°$, and giving to α the successive values, 5°, 10°, 15° . . . 40°, we have worked out the quantities y_1, y_0, and y_2, for a weight at such distances from the crown, and

quantities are given in the first portion of the following it being understood that the weights are here placed on ..ft of the crown to correspond with our figure: —

α.	y_1.	y_0.	y_2.	H.	P_1.	P_2.
0°	.0449 r	.3587 r	.0449 r	1.126 W	.5 W	.5 W
5	.0252	.3585	.0607	1.095	.596	.411
10	.0001	.3578	.0735	1.007	.683	.325
15	—.0341	.3569	.0842	0.866	.760	.244
20	—.0817	.3555	.0930	0.690	.830	.172
25	—.1536	.3537	.1012	0.498	.890	.111
30	—.2730	.3515	.1078	0.311	.939	.063
35	—.5137	.3487	.1142	0.150	.972	.027
40	—1.2407	.3470	.1183	0.040	.993	.007

These values of y_1, y_0, and y_2, have been plotted on the arch of Fig. 44, and the several stress diagrams have been drawn on a vertical line which represents W. From this figure the amounts of H and of the vertical components of the abutment reactions for a load W at successive points can be scaled off, and thus we obtain the last three columns of the above table. H, P_1, and P_2, can also be easily calculated by the formulæ of § 63, if we make $c = r \sin \beta$, and $b = r \sin \alpha$.

Having divided the centre line C A of the arch-ring of Fig. 45 at points C, F, G, &c., distant five degrees from one another, the weight to be concentrated at each of these loaded points is next computed, for an arch one foot thick, perpendicular to the plane of the paper, by scaling the area between the dotted ordinates, marked on the horizontal line, and placed midway between the points of division, and multiplying this area by the weight of a cubic foot of masonry, here assumed at 150 pounds. The weights at the several points, to the nearest hundred pounds, will then be

$$C = 7{,}500,\quad F = 7{,}600,\quad G = 8{,}400,\quad I = 9{,}600,\quad K = 11{,}100,$$
$$L = 12{,}800,\quad N = 14{,}600,\quad O = 16{,}600,\quad P = 19{,}300 \text{ lbs.};$$

making the weight of the half-arch (when we take one-half of the load at C, and add 9,800 pounds for the load at A), = 113,-450 pounds.

Calculate H for steady load by multiplying each co-efficient of H in the table above by its W in pounds just ascertained, and adding all the results for both halves of the arch. The work in detail is below. As the two halves of the arch are alike, we add up the column for H, add in again all but the amount for the load at the crown, and have H′ for the entire arch. Each vertical reaction will equal the weight of the half arch.

To find the ordinate $y_1' = y_2'$, for the combined weights, multiply each H by its y_1, add the products, and divide by H′. As, for each weight on one half of the arch, there will be a corresponding and equal weight on the other half, it will shc the process to add y_1 and y_2 together for each point on one-ı of the rib, *except the centre one* at C.

			W.	H.	$y_1 + y_2$.	M_1.	
C.	0°	1.126 ×	7,500 =	8,445 lbs.	.045 r +	380.0 r lbs.	
F.	5	1.095	7,600	8,322	.086	715 7	
G.	10	1.007	8,400	8,459	.074	626.0	
I.	15	0.866	9,600	8,314	.050	415.7	
K.	20	0.690	11,100	7,659	.011	84.2	
L.	25	0.498	12,800	6,374	—.053		—337.8 r lbs.
N.	30	0 311	14,000	4,541	—.165		749.2
O.	35	0.150	16,000	2,490	—.400		996.0
P.	40	0.040	19,300	772	—1.123		867.0
				55,376 lbs.		+2,221.6	—2,950.0
				46,931		—2,950.0	
			H′ =	102,307 lbs.		) —728.4 × 100 (	—.712 ft. = y_1'.

144. **Equilibrium Curve for Steady Load.**—Plot the weights of the above table on a vertical line from 1′ to 10′, lay

from the middle of 1′–2′ to 0′, and, starting at 0.71 feet , A, draw an equilibrium polygon with its sides succes- ly parallel to the lines which would radiate from 0′. This polygon will run quite close to the centre line, crossing it twice between A and C, and passing 0.4 feet below it at the crown. In any actual example the whole polygon should be drawn, as its accuracy will be proved by its striking the ordinate from B at the proper distance. If this arch were never to be subjected to any other than a steady load, or should the travelling load always be light, voussoirs of moderate depth would contain this polygon within their middle third. The true equilibrium curve will pass through the angles of the polygon just drawn.

145. **Calculations for Rolling Load.**—But, as we stated that a line of railroad was to be carried over this arch, let us suppose that the rolling load of one ton and a half per foot of track, or 3,000 pounds, is distributed over the ten feet of width of the arch; the moving load will then amount to 300 pounds per foot of span on the rib of our figure. The sleepers, the filling over the rib, and the bond of the arch-stones, will distribute any concentrated load over a considerable area.

At the crown of the arch the curve already drawn falls somewhat below the centre line. Upon inspecting Fig. 44 we see that six of the polygons there drawn pass below the crown of the rib. If, therefore, we place upon the stone arch a rolling load which covers six points of division from each abutment, that is, from Q to R on one side, and a corresponding distance on the other half arch, this distribution of load, if a practicable one under the usual method of running trains, will cause the greatest deviation of the equilibrium curve at the crown C.

To draw the polygon for this rolling load alone: first multiply each horizontal distance belonging to I, K, L, &c., by 300 pounds, to obtain the concentrated load on each point; then multiply by the proper co-efficients of H already obtained; sum the products, and double the results for both halves of the arch; multiply each H by its y_1 and y_2; divide the algebraic sums of these products by H″. The operations are carried out below.

		W.		H.	$y_1 + y_2$.		
I.	8.4	2,520 ×	.866 =	2,182	.050	+ 109.1 *r* lbs.	
K.	8.2	2,460	.690	1,697	.011	18.7	
L.	7.9	2,370	.498	1,180	—.053		—62.6 *r* lbs.
N.	7.5	2,250	.311	700.	—165		115.4
O.	7.1	2,130	.150	320.	—400		127.8
P.	6.7	2,010	.040	80	1.123		90.0
		13,740		6,159		+ 127.8	— 395.8

H″ = 12,318) — 268.0 × 100 (—2.2 ft. = y_1''.

Lay off the loads for one-half of the rib on a vertical line from 4″ to 10″; make 4″– 0″ = H″; and, laying off $y_1'' =$ - feet, at A, draw the polygon which passes horizontally below C at a distance, by scale, of 2.3 feet.

146. **Increase of Bending Moment at Crown; Required Depth of Keystone.** — We can now find how much this added load increases the negative bending moment at the crown of the rib, or how much it causes the equilibrium curve to move inwards. If we multiply H′ and H″ by the ordinates to their respective curves at the crown, which ordinates are 0.4 feet and 2.3 feet, as lately stated, and add the products, we shall obtain the existing moment at the crown, and, upon dividing by H′ + H″, we get the ordinate from the centre line at C to the curve for the combined loads. It is worthy of note how little effect the rolling load produces, owing to the great thrust of the masonry itself.

102,307 × 0.4 = 40,922.8 ft. lbs.
12,318 × 2.3 = 28,331.4
114,625)69,254.2
Ordinate at C = 0.60 ft.

In order that this deviation of 0.6 feet from the middle of the joint shall not bring the equilibrium curve outside of the middle third, the keystone and adjoining voussoirs must not be less than $0.6 \times 6 = 3.6$ feet deep. The greatest intensity of pressure, found at the inner edge, will then be twice the mean intensity of pressure, or $2\ [114{,}625 \div (3.6 \times 144)] = 442$ pounds per square inch, giving a factor of safety against crushing of about ten, for good limestone or sandstone. If the depth of the joint be increased to *four* feet, the greatest intensity of pressure at the inner edge will be reduced to $\frac{4+3.6}{4} \cdot \frac{114{,}625}{4 \times 144} = 378$ lbs. per square inch.

147. **Increase of Bending Moment at Haunch.** — The steady load curve deviates outwardly from the centre line the greatest distance, 0.5 feet, at L. Fig. 44 again shows that a rolling load from Q to R of Fig. 45 will increase this deviation to the greatest extent. The value of the horizontal thrust, H‴, for this load, will be seen, from the table of § 145, to be 6,159 pounds. Multiplying the same values of H by the then existing values of y_1, and proceeding as usual, we shall obtain y_1'''. If the total M_1 of this table is subtracted from

	H.	y_1.	M_1.	W.		P_2.
I	2,182 ×	−.034 =	− 74.2 *r* lbs.	2,520 ×	.244 =	614.9 lbs.
K	1,697	−.082	−139.2	2,460	.172	423.1
L	1,180	−.154	−181.8	2,370	.111	263.1
N	700	−.273	−191.0	2,250	.063	141.7
O	320	−.514	−164.2	2,130	.027	57.5
P	80	−1.241	− 99.3	2,010	.007	14 1
	6,159 lbs.		) −849.7 × 100 (−13 8 ft. = y_1'''.			$P_2''' = 1{,}514.4$ lbs.
			) +581.7 × 100 (+ 9.4 ft. = y_2'''.			

that of the table in § 145, we shall obtain the moment at B, and thence find y_2'''. To obtain the vertical component of one reaction, multiply each load by the proper co-efficient of P_1 or P_2, given in § 143. Since P_2''' is 1,514.4 pounds, lay this amount off from 4″, draw H‴ to 0‴, and plotting $-y_1'''$ at A, and

$+ y_2'''$ at B, draw that equilibrium polygon which passes 7.1 feet above L.

By the same process as before, we find that the equilibrium curve for the steady load, combined with these six loads on the left side of the arch, will be displaced from the centre line vertically at L 0.875 feet. The depth of the arch-ring at this point should, therefore, not be less, vertically, than 5.25, or,

102,307	× 0.5	=	51,153.5 ft. lbs.
6,159	× 7 1	=	43,728 9
108,466			)94,882.4
Ordinate at L =			0.875 ft.

measuring normally, than $5.25 \times \cos 25° = 5.25 \times 0.9063 =$ 4.76 feet.

148. **Influence of an Additional Load.** — When it is noticed that an additional load on the point G will cause the greatest positive moment at K, it may be suspected tha seven loads will cause a greater deviation at K than the o just found at L. To ascertain the fact, we may dispense with any new polygon by proceeding as follows: The new load G will be $8.6 \times 300 = 2,580$ pounds. H for this point, being 1.007 W, will equal $2,580 \times 1.007 = 2,598$ pounds. By scale, in Fig. 44, the ordinate from the proper polygon to the arch at the point K is $.017\,r = 1.7$ feet. The ordinates to the curves already drawn in Fig. 45 being scaled at K, the annexed computation is readily made, and the quotient is seen to be less than the amount at L. Kindred steps might be taken for any point.

102,307	× 0.35	=	35,807.4 ft. lbs.
6,159	× 8.10	=	49,887.9
2,598	× 1.70	=	4,416.6
111,064			90,111.9

149. **Increase of Bending Moment at Springing; Maximum H.** — The remaining point of maximum deviation of the curve for steady load is at the springing A, where we have found it to be .71 feet. As the same six loads from Q to R will be seen, from Fig. 44, to produce the maximum effect at A, the polygons are already drawn to our hand, and the moments at the springing point are seen in the respective tables. There-

fore the ordinate at A is 1.45 feet, and the normal displacement is $1.45 \times \cos 45° = 1.45 \times .707 = 1.03$ feet. The necessary depth for this joint will be 6.2 feet. If the amount of P_1 from rolling load, 12,226 pounds, is laid off below 10′, and H‴, 6,159 pounds, is plotted to the right of 0′, the line connecting the two points thus found will be the thrust at A, and, from its projection on a line inclined at 45°, we get 158,000 pounds for the direct thrust at A. The maximum intensity of compression on this joint will be at the inner edge, and will be 2 [158,000 ÷ (6.2 × 144)] = 354 pounds per square inch.

$$
\begin{array}{llll}
102{,}307 \times & .71 & = & 72{,}840 \\
\underline{6{,}159} \times & 13.8 & = & \underline{84{,}970} \\
108{,}466 & & &)157{,}810 \\
\text{Ordinate at A} & & = & 1.45 \text{ ft.}
\end{array}
$$

The maximum value of H will occur when the rolling load covers the whole bridge. If the amounts of H for the points have not yet been loaded are computed, the horizontal ... for a complete travelling load will be found to be 26,206 ...nds. The equilibrium curve for such a load will be a para-...ola; the ordinates y_1 and y_2 will be 1.19 feet, and the curve will pass the crown at a distance of + 0.5 feet vertically. As this parabola, when drawn if desired, will be found to lie at most points on the opposite side of the centre line from the curve for steady load, the effect of a complete rolling load will be to bring the arch quite near to actual equilibrium. The deviation at the crown will be reduced to — 0.2 feet, and, as the total thrust will then be 128,513 pounds, the greatest intensity of compression at that section, for a four-foot voussoir, will be $\frac{4 + 1.2}{4} \cdot \frac{128{,}513}{4 \times 144} = 290$ lbs. on the square inch. We have now examined in detail all of the critical points of this arch.

150. **Final Dimensions of Arch.** — The arch-ring was assumed, at the start, to be five feet deep. It is apparent, from our investigation and the conditions imposed, that this depth is greater than is necessary for the larger part of the arch, but is less than is required near the springings. For a travelling load of somewhat less intensity, a ring having a uniform depth of

five feet will be entirely satisfactory. Guided by these results, we may redistribute the steady load in the spandrels so as to bring the equilibrium curve for that load nearer the centre line at the springings. Another trial will probably accomplish the desired end, and the above curves for rolling load can be used anew. Otherwise, the arch-ring may be made four feet deep at the crown, and six feet and a half deep at the apparent springings, as shown on the right half of Fig. 45, and in that case the curves which have been discussed will lie within the middle third of the rib. Although the formulæ for the circular arch were derived upon the assumption that the rib was of constant thickness, the deviation which we suggest will hardly be of serious consequence. The tenacity of the cement, and the greater or less resisting power of the material immediately in contact with the ring, will sufficiently provide for all contingencies. We have therefore drawn this form as the final determined shape of the arch-ring, the centre line being undisturbed, and the radii of the intrados and extrados being about 95 feet and 104 feet respectively. One must remember, that, as the ring has been altered from a uniform depth of five feet, care must be taken to put a little more filling at the crown, and less at the springing, in order that the distribution of the steady load may be unchanged.

151. **General Remarks.**—If the exterior spandrel wall is massive, a separate equilibrium curve may be required for that portion of the ring which carries the wall: such portion will be subjected to a steady load equal to the weight of the wall, but need not be considered as carrying any travelling load. It was not our purpose to enter into the subject of the construction of stone arches, but to show the method of finding the forces which act on a given or assumed rib. Two or three matters, however, will be briefly referred to. If, at any point, the direction of the resultant pressure makes a considerable angle with the tangent to the centre line of the ring, the two voussoirs having a joint at that place might slip on one another if the joint were radial. No joint should deviate very far from a

plane perpendicular to the pressure. Generally this angle of deviation is too small to be of practical importance, and the joints are made radial or normal to the intrados.

In case several arches are built in a series, it is well to so proportion the span and rise of each, that the horizontal thrusts from steady load may nearly balance one another, as we shall then avoid a disturbance of one arch by the other, and can carry the arches on reasonably slender piers. If one arch has more thrust than the other, and the pier between the two yields, we have a change of span, like that due to temperature, so far as its treatment goes; and its effect upon the arches can therefore be determined.

As we know the direction, amount, and point of application of the thrust at the springing, we can easily construct the line of thrust, or equilibrium curve for the abutment, by combining the weight of the abutment and of the mass of masonry immediately above it with this thrust at the springing, the weight of the masonry just above this point being first compounded, and then the weights of successive portions of the abutment. Hence the required thickness of the abutment is ascertained.

152. **Exaggeration of Vertical Scale.**—Since some of the equilibrium curves may run quite close to the centre line, especially the one for steady load, it may improve the accuracy of measurement of the ordinates or displacements to exaggerate the vertical scale of the drawing. In this case, since all vertical lines will be increased in length, the load lines of the stress diagrams must be laid off with the same proportion to those which represent H. This suggestion immediately opens the question of the possibility of treating elliptic ribs.

153. **Elliptic Arch.**—If we refer to the original equations of condition for any rib, viz., Σ E F . D E $=0$, Σ E F $=0$, and Σ E F . D B $=0$, it is apparent, that, if all the ordinates D E and E F of a circular rib are multiplied or divided by any given quantity, the summations indicated above will still equal zero, and that the ordinates y_1, y_0, and y_2, thus determined, will apply to an elliptic rib whose semi-axes are obtained from the radius

of a circular rib by the same multiplication or division. This fact is easily seen by reference to Fig. 43. Here are drawn a semicircular rib and two elliptic ribs, of the same span, but differing in height; one having one-half the rise, and the other one and one-fourth the rise, of the semicircle. We will suggest, that, to find points on an ellipse, a simple way is to draw from the same centre two semicircles whose radii are the semi-axes of the ellipse; then prolong any radius; from the point where it cuts one circle draw a horizontal line, and, from the point where it cuts the second circle, draw a vertical line; the intersection of the lines last drawn will be one of the desired points. This construction is seen in the figure, and, as one of the circles is needed subsequently, the method is convenient.

154. **Example** —Taking a load at 30° from the crown of the semicircular rib as an example, we find, by turning to the table of § 99, that $y_1 = .360\,r$, $y_0 = 1.298\,r$, and $y_2 = .011\,r$: the polygon is plotted on the semicircle of the figure. In the upper sketch every ordinate for the ellipse being one-half of the corresponding ordinate for the circle from which it is projected, we have simply to substitute the semi-axis $a = \frac{1}{2}\,r$ for r, and we have $y_1 = .360\,a$, $y_0 = 1.298\,a$, &c. The equilibrium polygon may then be drawn, and it is apparent to the eye that it satisfies the imposed conditions alluded to in the last section. Similarly, for the other ellipse, we write a for r, in that way multiplying the ordinates of the semicircle by 1.25.

It is evident that the points of contraflexure are unchanged in horizontal position, as is also the *horizontal* distance of the imposed load from the crown; but the symbol $\alpha = 30°$ of the example has no significance in the ellipse as denoting the *angular* distance of the load from the crown. We must, in place of such notation, either draw the semicircle which has the *span* for a diameter, and work from that, as has here been done, or else for α read $r \sin \alpha$, where r equals horizontal semi-axis of the ellipse, and lay off the distance from the centre on the diameter to locate the foot of y_0. A segment of a semi-ellipse can be treated exactly as a segmental circular rib is treated: it will be

such a load, when the span, rise, and depth are all given. In the same way that an ellipse is derived by projection from a circle, a curve, called a transformed catenary, can be projected from a catenary, and will be in perfect equilibrium under the desired or prescribed wall. While some of the quantities used are derived by mathematical analysis, which we will not insert here, the accuracy of these quantities can be verified from the diagram.

Let it be desired to find the form of the arch, of half span P Q, which shall be in equilibrium under masonry whose depth at the crown shall be S P, and at the springing R Q. It is understood that the arch will be inverted from this figure, and it will be seen that this type of arch may be applied to any span and rise. Let P Q $= c$, P S $= h_0$, Q R $= h_1$, P O $= m$, and Q A $= y_1$. The first step will be to find the value of P O, and thus determine the original catenary. This will be done by solving the equation

$$m = \frac{c}{2.30158 \times \log\left(\frac{h_1}{h_0} + \sqrt{\frac{h_1^2}{h_0^2} - 1}\right)};$$

where *log.* denotes the common logarithm of the quantity in the parenthesis. Let the half-span be 30 feet, the rise 8 feet, and the depth of load at the crown 2 feet; then is h_1 10 feet, and the above expression becomes

$$m = \frac{30}{2.30158 \times \log(5 + \sqrt{24})} = \frac{30}{2.29242} = 13.09 \text{ ft.}$$

Then by proportion

$$h_0 : m = h_1 : y_1, \quad \text{or} \quad y_1 = \frac{m\,h_1}{h_0} = 13.09 \times 5 = 65.45 \text{ ft.}$$

We next obtain from the following formula, the length of the catenary,

$$s = \sqrt{(y_1^2 - m^2)} = \sqrt{(65.45^2 - 13.09^2)} = 64.1 \text{ ft.},$$

and

$$\frac{P_1}{H} = \frac{s}{m} = \frac{64.1}{13.09} = 4.9.$$

We may now proceed to draw the catenary between the points A and O. Any length of load line may be laid off, and H then drawn of the proper proportionate amount just found. But, if preferred, P_1 may be made equal to the weight on the catenary, which will be the area between the curve and the directrix multiplied by the weight of a cubic foot of masonry. The area can be proved equal to $m\ s$, or the product of P O by the length of the curve just found. Divide the load line into a certain number of equal parts, and divide s by the same number. Then proceed with the construction of § 156.

158. **Construction.** — The transformed catenary must be a projection of the catenary so drawn, and the load and load line will be reduced in the same proportion. To save the trouble of redividing the load line, multiply 1–0 by the ratio $m \div h_0$; that is, enlarge the scale of the stress diagram, and lay off that distance from 1 to 0′. Radiating lines from 0′ to the old points of division will be parallel to those which might be drawn from 0 to new points of division; therefore, starting from R, draw the curve R S by making its sides parallel to lines radiating from 0′, and bringing the points B′, C′, D′, &c., vertically below B, C, D, &c. But it must be remembered that H in the new curve is the same in amount as H in the old one, while P_1, the vertical component of the reaction, is reduced in the ratio just referred to. The rib need only be deep enough to have strength to resist the thrust. Fig. 42 shows the arch in an erect position.

159. **Many-centred Arch.** — If it is wished to lay out an approximation to the transformed catenary, composed of arcs of circles, draw normals at the middle points of the successive sides of our construction, and, to get them accurately, make them perpendicular to the radiating lines of the stress diagram. Prolong them until they intersect one another, and, on or near the curve which can be sketched through those intersections, select as many centres as may be desired for the circular arcs. Thus arches of three, five, or seven centres may be drawn, which will be good approximations to the transformed catenary.

in structures will be appropriate. The introduction of three hinges will do away with these sources of error. This type of stiffening truss will be discussed further in connection with the one which follows.

162. **Horizontal Girder.**—It is much more common to employ a horizontal truss or girder, as shown in Fig. 47, to stiffen the suspension-bridge. If we note that the office of the arch or inverted arch is twofold,—first to resist the direct stress, and, second, to resist the bending moments at successive sections, —we see that the horizontal girder of this figure will be subject to the same bending moments at similar sections as the inverted arch or braced rib of Fig. 46, while the chain will here carry the direct stress, which in the former case was also resisted by the rib.

If the truss is hinged at the middle as well as at the abutments, it comes under the class of Chap. II.; and the effect of one or more loads is easily determined. We may draw Fig. 48, if desired, and find by inspection the extent of rolling load required to produce the maximum bending moment of either kind at any point. See § 32. Thus, at one-fourth the span from one abutment, the maximum bending moment of one kind occurs when the rolling load covers four-tenths of the span on the same side; and the maximum bending moment of the opposite kind, when the rolling load covers the other six-tenths of the span. The maximum moment at a point near the abutment is found when the head of the load is at one-third the span from that abutment. These values are easily deduced by finding the horizontal distance of the point of intersection D, in Fig. 48, on A F, of that line, which, starting from B, passes through E, the extremity of a certain ordinate. Those authors who make maximum bending moments at all points occur, for a stiffening girder hinged at ends and middle, when the half-span is covered, are in error. The shear diagrams are constructed as explained in the earlier chapters. The construction for normal shear will be applicable to Fig. 46, and the vertical shear diagram to the stiffening truss of Fig. 47.

163. Distribution of Rolling Load between Cable and Truss. — It may be well to call more particular attention to the distribution of the rolling load between the truss and cable of Fig. 47, and the way in which bending moments are caused in the unloaded portion of the horizontal girder. If the bridge is unequally loaded, and no stiffening appliances are used, a distortion is produced, as explained in the first section of this chapter. When a weight W is applied on a suspension-bridge of half-span c, at any point distant b from the middle hinge, we know, in the first place, that the *total* reaction at A, Fig. 47, the end farthest from the weight, is $W\frac{c-b}{2c}$, and at B is $W\frac{c+b}{2c}$; and, in the second place, as there can be no shear in the cable, we see, from the equilibrium polygon of Fig. 48, and the lines 0–4 and 0–8, drawn in the stress diagram parallel to the tangents to the cable at the tops of the towers, that $5\text{–}4 : H = 2k : c$, or $5\text{–}4 = \frac{2k}{c}H$. By § 23, $H = \frac{c-b}{2k}W$; therefore the amount of vertical force combined with H of the cable is $W\frac{c-b}{c}$. Hence at A and at B the cable itself produces a reaction of $W\frac{c-b}{c}$, the balance of the reaction comes from the truss; the reaction of the truss at A will therefore be $-W\frac{c-b}{2c}$, and at B will be $W\left(\frac{c+b}{2c} - \frac{c-b}{c}\right) = W\frac{3b-c}{2c}$. This reaction also will be negative when b is less than $\frac{1}{3}c$. Such is the case in Fig. 48, for the polygon A D B; and we have a corroboration in the negative bending moments near each end.

As the vertical force at A or B from the cable is the load on the half-span of the cable, and this load must be uniformly distributed horizontally to keep the cable in its curve, the intensity of vertical pull exerted between the cable and the rods per horizontal foot is found by dividing the above force by the half-span: hence it is $W\frac{c-b}{c^2}$. This will be the *upward* pull on the girder per horizontal foot at all points and the cause of the bending moments. Of course at the point of application of W the resultant force acts downward. The action of a continuous load over a greater or less portion of the girder will follow the same law; and we shall have downward forces on the loaded portion of the girder equal to the difference between the imposed load and the pull of the vertical rods, and upward forces on the unloaded portion.

It is convenient to notice that the amount of W carried by either half of the cable is that portion which would be carried by the middle hinge if the half-girder alone supported W. As the girder reaction at the farther abutment is one-half of this amount, and the half-girder on the unloaded side is subjected to a uniform upward force, the shear on the middle hinge will also be one-half of this amount, or $W\frac{c-b}{2c}$. The shear diagram is given in Fig. 48. For any extent of load it will now be easy to find the amount carried by the cable; for we have only to calculate the portion which would come upon the middle hinge, were that a point of support of a simple truss of span c, and this portion will be the load on the half-cable.

164. **Comparison of Inverted Arch and Horizontal Girder.** — All statements in regard to the horizontal stiffening girder are equally true of the two parallel chains with bracing. While, in the bridge formed of cable and horizontal girder, the girder resists bending moments, and the chain takes up the direct stress, in the latter case the cables have to resist both moment and direct stress. But the maximum direct stress at any section, half of which is borne by each cable, occurs when the bridge is fully loaded: the maximum bending moment is found with a partial load, at which time the direct stress is less. Hence less material is theoretically required for the cables and truss of the type of Fig. 46 than for one like Fig. 47, — perhaps three-fourths as much. The introduction of the middle hinge in the axis of the rib of Fig. 46, with connections of sufficient strength to transmit the cable stresses, is attended with a little difficulty, which does not exist in the other case.

The three-hinged girder or rib may have the third hinge removed from the middle towards one end, as shown in Fig. 50, where one portion of the girder takes the form of a short link, extending to the first suspending rod. The same device may be introduced in an arch. The effect on the equilibrium polygon and the derived quantities may at once be seen.

165. **Horizontal Stiffening Girder hinged at Ends only.** —In case the middle hinge is omitted, the girder will be exposed to bending moments, as explained in Chap. III. Here, again, an inspection of Fig. 8 will show the extent of load required to produce maximum M of either kind; and an examination of the table of bending moments in the chapter referred to will show that an absolute maximum M occurs at one-fourth of the span from either abutment for a continuous load extending from one end to a point distant 0.43 of the span from the end nearer to the point of maximum M. Its amount is about $\frac{1}{7.5}$, or .133 of the maximum moment at the middle of an unassisted girder of the entire span. The stretching of the cables on both sides of the towers impairs the accuracy of these deductions. With a truss hinged at the middle, the sagging of the main cable, as well as the change of temperature, is of little consequence. From the value of Y_1, § 50, it is evident that $\frac{5}{32}(1 - n^2)(5 - n^2)W$ is carried by either half-chain, and this quantity divided by c will give the intensity of upward pull on the truss from a load W at one point. The above amount is again that which would be carried to the point of contraflexure of the truss, if that were the point of support of the unassisted truss, and the truss were discontinuous over the support. (Compare Rankine's "Applied Mechanics," p. 375, *note.*)

If the ends of the girder are fixed in direction, we have the case of Chap. IV. Enough has been said to plainly indicate the treatment.

166. **Stiffening Girder of Varying Depth.** — Returning anew to the case of the stiffening girder with three hinges, it is

evident, that if the girder has a variable depth, greatest at the points of maximum bending moment, the stresses in the flanges or chords will be diminished proportionally, with an economy of material. If, at the same time, the girder is itself the suspension cable, we can so adjust the depth, that the flange stresses for a partial load shall never exceed those arising from an entire load. Modifications having this end more or less in view have been suggested and carried out. Let us first draw, in Fig. 49, the equilibrium curve for a rolling load alone over half the span. While this curve will not give maximum bending moments, it will not differ greatly from the curves of maximum M, and it offers a very convenient and sufficiently accurate basis of comparison. Its form will be a straight line over the unloaded half of the span, and a parabola tangent to that line for the remaining portion. As the tangent at the abutment end of this parabola meets the tangent from the other end in the vertical through the centre of gravity of the load, the tangent A D is at once drawn. Draw the chord A C. The parabola cuts the middle vertical ordinate E D from the chord A C at its middle point F. If the height of the original parabola of the cable is k, the ordinate at one-fourth the span is $\frac{3}{4}k$. G D $= \frac{3}{2}k$; G E $= \frac{1}{2}k$; therefore E D $= k$; E F $= \frac{1}{2}k$; and F G $= k$. Hence the remaining ordinate for bending moment at one-fourth the span is $\frac{1}{4}k$ on either side, and of opposite signs.

167. **Ead's Arch, or Lenticular Stiffening Girder.**—If the two half-ribs of the arch of Fig. 51, or of the stiffened suspension-bridge, are each made of two equal parabolas, the outer ones being the continuous equilibrium curve for a complete load, the vertical depth of the semi-girders at their middle sections E and F will be one-half the rise or height, k. Let us denote the horizontal thrust or tension from steady load w by H; that from a full rolling load w', by H′. The horizontal stress due to a rolling load extending from one abutment over half the span will be $\frac{1}{2}$ H′; for a similar load over the other half-span must give an equal stress, and both combined must equal H′. When the above bridge is fully covered with mov-

ing load. the equilibrium curve will coincide with the continuous curve, and the stress at each section of the main cable will be that due to $H + H'$. The auxiliary ribs and bracing will experience no stress. When the bridge is half loaded, say from C to B, the equilibrium polygon for rolling load will be the one sketched in our figure; it passes at I, $\frac{1}{4} k$ below the main cable at D, and through the middle or axis of the truss A C. The horizontal component of the stress at D, due to $\frac{1}{2} H'$ at I, is, from the equation of moments about E, $\frac{3}{4} H'$; that is, $\frac{1}{2} H' . \frac{3}{4} k =$ hor. comp. at D $\times \frac{1}{2} k$. Taking moments about D, $\frac{1}{2} H' . \frac{1}{4} k = -$ hor. comp. at E $\times \frac{1}{2} k$; or horizontal component at E is $- \frac{1}{4} H'$. At F and G the horizontal component is, in each member, $\frac{1}{4} H'$. The minus-sign denotes opposite stress, here compression; in the arch, tension. We may therefore write the following table of cases:

Horizontal component of stress at .	E	D	F	G.
With steady load only	0	H	0	H,
" " and one-half rolling load	$-\frac{1}{4}H'$	$H+\frac{3}{4}H'$	$+\frac{1}{4}H'$	$H+\frac{1}{4}H'$,
" " " complete " "	0	$H+H'$	0	$H+H'$.

Since F and G change places with E and D for a load on the other half-span, we see that the lower member, or main cable, experiences a horizontal component which fluctuates from H to $H + H'$, always tension; while the auxiliary rib has a stress whose horizontal component ranges between $\frac{1}{4} H'$, tension, and $\frac{1}{4} H'$, compression. The bracing will undergo no stress from a full load. The stress in the bracing for partial loads may be worked out by the method of the previous chapters for finding the amount of shear remaining after subtracting the vertical components for the two cables at a section, by the method of Part II., "Bridges," Chap. V., or by drawing stress diagrams as given in Part I., "Roofs."

As the parabola through I is a *projection* of that through D, the above deductions for the points D and E are true for the other points of the girder. Although, as pointed out in § 162, the bending moments are a little greater for loads which cover not quite half the span, it is evident that the horizontal component of the stress in the main cable can never exceed H + H′, and in the counter-rib will but slightly exceed ± ½ H′. This form of arch was designed and patented by James B. Eads: a paper upon it by him may be found in the "Transactions of the American Society of Civil Engineers," vol. iii., No. 6, October, 1874.

168. **Bowstring Stiffening Girder.** — If the auxiliary members connecting the hinges A, C, and B, Fig. 52, are straight, we have a variation in the method of stiffening and a change in the stresses. The equilibrium curve A F C I B, for a rolling load over one-half the span, is also drawn here, coinciding with A C, and passing through I, ½ *k* below D. The steady load will be entirely carried by the main cable as before, as will also a complete rolling load. The half rolling load, being entirely supported on the left by A F C, will cause in that member a tension whose horizontal component is ½ H′; a horizontal tension in D, of H′, and a horizontal compression in E, of ½ H′, as is found by similar equations of moments to those in the last section. There results, then, for this type the following cases: —

Horizontal component of stress at . .	E	D	F	G,
With steady load only	0	H	0	H,
" " " and one-half rolling load	−½ H′	H + H′	+ ½ H′	H,
" " " " complete " "	0	H + H′	0	H + H′.

The stress on the main cables will be very slightly increased for some partial loads, as shown before. The increase will, however, be small, for the direct stress is decreased at the time the bending moment is increased; so that the absolute maximum may be called H + H′ without any error of importance. The stress in the straight stiffening rib ranges from a tension of ½ H′ to a compression of ½ H′. While the member A C or C B has to resist double the force of the preceding case, and that force also completely reversed for a moving load over one-half of the bridge, the unbraced lengths are shorter than in Fig. 51, the construction of a straight member is simpler, and the web members are only one-half as long: the cost may therefore be sufficiently influenced to cause this design to commend itself more to the practical builder than does the former. A notable example of this type is the Point Bridge at Pittsburgh, Penn., eight hundred feet span, built by the American Bridge Company of Chicago, in 1876.

169. **Fidler's Stiffened Suspension-Bridge.** — Again, let us conceive of two cables, A F C D B and B E C G A, Fig. 53, each separately subject to, and in equilibrium under, a rolling load over one-half the span, and then let their places be taken by the two girders shown. A C and C B will be straight, as in the last figure; A G C and C D B will be parabolas, each tangent at C to the chord of the other; and the equilibrium curve for a complete load will pass through the middle of each truss, as shown by the dotted line. These trusses are, therefore, of the form of Fig. 52; but they have a depth equal to that of the trusses of Fig. 51. The horizontal component H, of steady load, and H′, of complete rolling load, will be carried equally by both members of each truss, $\frac{1}{2}$ H and $\frac{1}{2}$ H′ on each. A rolling load on the right half of the span will cause a horizontal tension of $\frac{1}{2}$ H′ at D and at F. We may, then, write, for this type,

Horizontal component of stress at	E	D	F	G.
With steady load only	$\frac{1}{2}$H	$\frac{1}{2}$H	$\frac{1}{2}$H	$\frac{1}{2}$H,
" " " and one-half rolling load	$\frac{1}{2}$H	$\frac{1}{2}$H+$\frac{1}{2}$H′	$\frac{1}{2}$H+$\frac{1}{2}$H′	$\frac{1}{2}$H,
with steady load and complete rolling load	$\frac{1}{2}$H+$\frac{1}{2}$H′	"	"	$\frac{1}{2}$H+$\frac{1}{2}$H′.

The stresses will, therefore, always be tension, and the horizontal component will vary in each member from $\frac{1}{2}$ H to $\frac{1}{2}$ (H + H′), a most favorable showing for the structure. The remark of § 162 in regard to maximum bending moments applies here also. The maximum stresses in the bracing can be worked up in the way thought most convenient. This type may also be analyzed as two inverted bowstring girders, a weight on one causing simply a tension in the tie of the other and an inclined reaction in its line at the middle hinge. Hence the investigation of the bowstring girder in Part II. may be applied here. A very interesting analytical discussion of the types of bridges and arches of this chapter may be found in "Engineering," vol. xx. for 1875, from the pen of Mr. T. Claxton Fidler, the inventor and patentee of the type discussed in this section.

170. **Ordish's Suspension-Bridge.**—Another stiffened suspension-bridge, in which the problem of resisting distortion from a partial load is solved in quite a different way, is what is known as Ordish's, shown in Fig. 55. The Albert Bridge over the Thames, at Chelsea, Eng., is of this type; and one of moderate span has been erected over the Pennsylvania Railroad, at 40th Street, Philadelphia. Here a certain initial stiffness is given to the platform itself, and it is then directly supported at several points from the tops of the towers. It is intended that the weight shall be entirely carried by the inclined ties. As those ties, from their length, would sag considerably under their own weight, a passing load would cause the roadway to move vertically; for an increased pull on a tie would tend to straighten it. They are, therefore, suspended, at the joints in the several bars which make up the ties, from a light cable, which is designed simply to carry the weight of the ties; and the suspending rods are so adjusted, that the ties shall be straight. No movement of the roadway of any importance can then take place. The analysis is very simple.

171. **Erect and Inverted Arch combined.**—The bridge over the Elbe, at Hamburg, one span of which is shown in Fig. 54, is a combination of the erect and inverted arch. This construction dispenses with abutments to withstand a thrust, as the thrust of the upper rib will at all times be balanced by the

tension of the lower rib. If the ribs are of equal stiffness, any load may be considered as divided equally between the two systems: if the ribs, while having the same curvature, are not alike in cross-section, the load will probably be distributed in the ratio of their moments of inertia. As the erect arch always tends to move away from its equilibrium curve, and the inverted arch to approach the equilibrium curve, the tangents at the abutment ends will move in the same direction, and therefore the structure should be treated as hinged at the ends, unless each flange is firmly bolted to the skew-back. If the structure is carried on columns or a pier, it appears to us that the ends cannot be rigid, and we judge that the two ribs will begin to turn about the middle of the depth without the introduction of a pivot or hinge.

The effect of temperature is annulled. Also the shortening of the erect arch under the direct compression being opposite to the extension of the inverted arch under the direct tension, the span will tend to remain unaltered; but the ribs themselves will be changed in form, one rib flattening as the other becomes more convex. If, in making such a design, the section of the arch is found to differ much from the section of the inverted rib, it will be well to calculate the relative deflections of the two ribs at the middle. The amount of load each will carry varies inversely as the deflection under equal loads, since they must deflect equally; and hence, if the arch is first designed of such shape, for the purpose of resisting compression, that it is stiffer or has less deflection than the chain, when each has one-half the load, the cross-section of the arch must be increased, and that of the chain may be diminished. This type of structure must not be confounded with a lenticular girder: the absence of bracing between the ribs makes them independent.

CHAPTER XI.

BENDING MOMENTS FROM CHANGE OF FORM.[1]

172. **Displacement from Bending Moments.**—It follows, from the fact that the arched rib moves away from the equilibrium polygon or curve, that the bending moments and chord stresses will have a slight tendency to increase. When the rib changes in shape, however, the equilibrium polygon must also move enough to still satisfy for the new form the equations of condition by which it was first established, and this movement will in some measure counteract the former. Besides, the equilibrium curve for steady load generally runs so close to the axis of the rib, that the change of shape from bending moments is very slight; and, even when the influence of rolling load is added, the increments of the bending moment ordinates are too small to be of material consequence.

The vertical displacement at any point E, Fig. 56, produced by any load, will be found, for the parabolic rib, by taking *area moments*, as explained in Part II., "Bridges," Chap. VI., or for the circular rib by summing the ordinates as usual along the rib. As was done in the treatment of beams, it will here be necessary to find the point D where the tangent to the rib in its new form is horizontal, which point will not be at the crown,

[1] Many of the deductions in this chapter are only intended as guides in practical construction, to indicate where, and to show approximately how much, additional stress may be anticipated from change of form. Exact results are not attempted.

--ept for symmetrical loads. D is then to be assumed momen-
dly as a fixed point, and the deflection or area moment of A nd E obtained with reference to it: the subtraction of the itter from the former gives the displacement of E relatively o the abutment A; that is, from the area moment between) and A subtract the area moment between D and E; and he remainder, when multiplied by H ÷ **E I**, will be the vertical lisplacement of E. As just stated, these displacements may be ıeglected.

173. **Displacement and Bending Moments from Compression.**—The thrust which exists at each section of the rib must, by its compression of the particles, cause a shortening of the rib, and, as the shorter rib must fit the same abutments, it is necessarily lowered at the crown. The resulting bending moments may be of consequence. So far as the rib retains sensibly its old form, parabolic or the segment of a circle, the equilibrium polygon is lowered proportionally to the sinking of the rib, as indicated in Fig. 57, in order to still satisfy the equations of condition; but, as the deflection v at the crown is very small compared with k, the alteration of the bending moment ordinates is very trifling. On the other hand, this lowering of the apex of the equilibrium polygon at once increases the value of H, offsetting the change first pointed out. This will be seen, also, from the values of M, § 44, into which k does not enter. The bending moments from the external load are therefore practically unaltered by the change of form.

To produce this change of form, however, or to bring the arch down to its new position, requires a change of inclination, and consequently a bending moment, at most points of the rib. The strains thus induced should be examined. Strictly accurate theoretical investigations for the different ribs cannot easily be made; but formulæ may be deduced which will serve all practical purposes.

174. **Parabolic Rib hinged at Ends.**—The parabolic rib which we have treated varies in cross-section, from the crown

to the springing, according to the secant of the inclination to the horizon, § 37; and, as the magnitude of the direct thrust for a complete uniform load varies in the same way, the intensity of direct compression per unit of cross-section arising from H will be constant, and every unit of length of arc will be shortened by that thrust the same amount, so that the arch will be altered as if exposed to a change of temperature. We will assume that the new form of the rib is still a parabola with a rise k' in place of k, but with the original span $2\,c$.

By definition, Part II., "Bridges," § 85, the modulus of elasticity E equals the intensity of stress divided by the shortening of a unit's length. Let the constant intensity of thrust equal the thrust at the crown H, divided by the cross-section at the crown A; let the compression of a unit's length equal the difference, $s-s'$, between the lengths of arc before and after compression divided by the original length s. Then

$$s - s' = \frac{H\,s}{A\,E}.$$

An approximate formula for the length of a parabolic arc is, in our usual notation, $s = 2\,c + \frac{4}{3}\frac{k^2}{c}$. The value of s' will be obtained by writing k' for k; then

$$s - s' = \frac{4}{3\,c}(k^2 - k'^2) = \frac{H\,s}{A\,E} = \frac{2\,H}{3\,A\,E}\cdot\frac{3\,c^2 + 2\,k^2}{c}.$$

As v, the deflection at the crown and the difference between k and k', is very small, we may write, without sensible error, $k - k' = v$, and $k + k' = 2\,k$; whence $k^2 - k'^2 = 2\,k\,v$, and we have

$$\frac{8}{3\,c}k\,v = \frac{2\,H}{3\,A\,E}\cdot\frac{3\,c^2 + 2\,k^2}{c}, \text{ or } v = \frac{H}{4\,A\,E}\cdot\frac{3\,c^2 + 2\,k^2}{k}.$$

It was proved, in § 36, that this rib deflected vertically like a horizontal beam of uniform section: hence to bring the arch down to its new position will create bending moments at all points such as would accompany the same deflection in a straight beam, supported at the ends, uniformly loaded, and of a cross-section equal to that of the rib at the crown. In Part II., "Bridges," § 95, we found, for a beam supported and loaded as above with w per foot,

$$v = \frac{5}{384} \cdot \frac{wl^4}{EI} = \frac{5wc^4}{24EI} = \frac{5M_0c^2}{12EI},$$

if M_0 is the bending moment at the middle. Equating these two values of v, we obtain

$$\frac{5M_0c^2}{12EI} = \frac{H}{4AE} \cdot \frac{3c^2 + 2k^2}{k},$$

or

$$M_0 = \frac{3IH(3c^2 + 2k^2)}{5Ac^2k},$$

.e additional positive bending moment at the crown of the arch, caused by its compression under the thrust H.

The bending moments at other points may then be taken to compare with those of the beam, that is, as the ordinates to the parabola, being $\frac{3}{4}M_0$ at the quarter-span.

175. **Remarks; Example.**—It will be noticed that **E** has disappeared from the expression for M_0: hence the bending moment will be the same, whether the material be iron, steel, or wood. As **I** contains A, and may be written nAh^2, Part II., "Bridges," § 86, n being a numerical factor, it is seen that the bending moment from deflection of the rib due to compression increases with the square of the depth of the rib, and, as $M \div h$ equals the flange stress, this stress will increase directly as the depth. To diminish the effect of change of form alone, employ a shallow rib.

If H = 20 tons, $c = 100$ feet or $l = 200$ feet, $k = 20$ feet, and $h = 2\frac{1}{2}$ feet, for a rib with two plate flanges and thin or open web, $I = 2\{\frac{1}{2}A \cdot (\frac{1}{2}h)^2\} = \frac{1}{4}Ah^2$, and

$$M_0 = \frac{3 \times 25 \times 20 \times 30{,}800}{5 \times 16 \times 10{,}000 \times 20} = 2.9 \text{ ft. tons at crown,}$$

giving 1.16 tons compression on upper flange, and an equal tension on lower flange.

176. Displacement from Change of Temperature. — The deflection produced by a fall of temperature in the parabolic rib hinged at the ends will be found by taking the area moment of the half parabolic segment, Fig. 16, from the crown to the springing about one abutment, and multiplying by $H \div EI$. Hence, as in Part II., "Bridges," § 95,

$$v_t = \frac{H_t}{EI} \cdot \tfrac{2}{3} ck \cdot \tfrac{5}{8} c = \tfrac{5}{12} \cdot \frac{H_t}{EI} \cdot c^2 k,$$

the deflection at the crown when the temperature falls, and the rise of the crown when the temperature rises. One may prefer to consider the rib in its new position as the proper curve from which to obtain the area moment. If it is assumed to still be a parabola with the rise k', we have

$$v = \tfrac{5}{12} \frac{H}{EI} c^2 k', \text{ and } k' = k \pm v.$$

Substitute this value of k', and v becomes

$$v = \frac{5 H c^2 k}{12 EI \mp 5 H c^2}.$$

This deflection is the result of the bending moments arising from H_t, and is not to be regarded in the light of the preceding section. The moments were computed in § 74. These moments will be slightly altered by the movement, as it shortens or lengthens the ordinates; but H_t will be changed in the opposite direction, reducing the actual modification of the moments. Since

$$H_t = \frac{15}{8} \cdot \frac{teEI}{k^2}, \quad v_t = \frac{25}{32} \cdot \frac{tec^2}{k},$$

a quantity independent of the cross-section of the rib, and, so far as the material is concerned, affected by the co-efficient of expansion only.

The bending moments due to the direct thrust, whether arising from a load or change of temperature, have been considered, as well as the resulting deflection. When the temperature rises, H_t is thrust, and in itself tends to shorten the rib, and thus reduce the above amount of rise due to expansion. The ratio of the two deflections will be

$$\frac{v}{v_t} = \frac{H_t}{4AE} \cdot \frac{3c^2 + 2k^2}{k} \div \frac{5}{12} \frac{H_t}{EI} c^2 k = \frac{3}{5} n h^2 \left(\frac{3}{k^2} + \frac{2}{c^2} \right).$$

In the example previously cited this ratio becomes

$$\frac{v}{v_t} = 0.6 \times \frac{25}{16} \left(\frac{3}{400} + \frac{2}{10{,}000} \right) = .0072,$$

a reduction of three-fourths of one per cent. When the temperature falls, H_t is a tension, and, in lengthening the rib, slightly reduces the deflection.

The deflection for a co-efficient of expansion of .000007 and a , of temperature of 30° will be, in our example of § 175,

$$v_t = \frac{25 \times 30 \times .000007 \times 10{,}000}{32 \times 20} = .082 \text{ ft.} = 1 \text{ inch.}$$

[The expansion or contraction of a straight bar may be conveniently stated as ¼ inch in one hundred feet for 30° F.] The theoretical movement of the rib at the crown for a range of 30° above and below the temperature at which it was constructed will therefore be two inches. The actual movement is generally less than theory would indicate, owing to gradual transition from one extreme to another, protection of some portions of the structure from extremes of temperature, as by shielding from the direct rays of the sun, &c., and, finally, imperfect freedom of motion.

177. **Initial Camber for Arch.**—It may be expedient to make the rib a little longer than the distance between the springings to compensate for the amount of compression which will arise from the steady load, or else to wedge up the springing points until the crown of the rib, when not under strain, shall be a distance v above its normal position: the rib will then, when in place and under its steady load, come down to the curve for which it is designed, and will be free from that portion of initial bending moment due to change of form from steady load. This will be true, because, in forcing the rib up,

we have introduced bending moments of the opposite kind to an equal amount. An additional allowance may be made for an ordinary travelling load. If the rib is to be made longer to offset the compression, find v, § 174, or H from steady load, and make the parabolic rib of a span $2\,c + u$ and a rise k, so that, when sprung into place on a span $2\,c$, it would rise to a height $k + v$, if it were not compressed at the same time.

Noticing, from § 174, that this compression acts like a fall of temperature in shortening the rib, we have, from § 74,

$$H_t = \frac{15}{8} \cdot \frac{\mathbf{E}\,I}{c\,k^2} \cdot t\,e\,c = \frac{15}{8} \cdot \frac{\mathbf{E}\,I}{c\,k^2} \cdot \frac{u}{2},$$

since u must equal $2\,t\,e\,c$. But $H_t = \frac{12}{5}\frac{\mathbf{E}\,I}{c^2 k}\,v$, by § 176, and, equating these two values, we get

$$\frac{15}{16} \cdot \frac{\mathbf{E}\,I}{c\,k^2} \cdot u = \frac{12}{5}\frac{\mathbf{E}\,I}{c^2 k}\,v,$$

or

$$u = \frac{64}{25} \cdot \frac{k}{c} \cdot v = \frac{16}{25} \cdot \frac{H}{A\,\mathbf{E}} \cdot \frac{3\,c^2 + 2\,k^2}{c}.$$

If, in our preceding example, A is eight square inches, and $\mathbf{E}$ is 24,000,000, u becomes half an inch.

178. **Parabolic Rib with Fixed Ends.** — In this case the deflection will naturally correspond with that of a beam of uniform section, uniformly loaded, and fixed at the ends, as will be seen by comparing the equilibrium curve of Fig. 17, where H from temperature alone acts, with that of such a beam. In Part II., "Bridges," § 99, and Fig. 47, we found that

$$v = \frac{w\,l^4}{384\,\mathbf{E}\,I} = \frac{w\,c^4}{24\,\mathbf{E}\,I} = \frac{M_0\,c^2}{4\,\mathbf{E}\,I},$$

if M_0 is the bending moment at the middle. Equating this value of v with the one found in § 174, we obtain

$$M_0 = \frac{I\,H\,(3\,c^2 + 2\,k^2)}{A\,c^2\,k}.$$

ending moment at the springings will be double this , and of the opposite sign.

leflection produced by a change of temperature will be y taking the area moment of the semi-segment of the ua already obtained in § 176, and subtracting the area ; of the rectangle whose height is $\frac{2}{3}k$ and base c.

$$v_t = \frac{H_t}{EI}\left(\tfrac{5}{12}c^2k - \tfrac{2}{3}ck \cdot \tfrac{1}{2}c\right) = \tfrac{1}{12}\frac{H_t}{EI}c^2k.$$

Applying the data of the previous example of § 175, we have

$$M_0 = \frac{25 \times 20 \times 30{,}800}{16 \times 10{,}000 \times 20} = 4.8 \text{ ft. tons at crown},$$

giving 1.92 tons, compression on upper flange and an equal tension on lower flange at crown, and 3.85 tons, tension on upper flange with an equal compression on lower flange, at either springing.

To find such additional length of span for the parabolic rib fixed at the ends, that, when compressed under steady load, it may have no bending moments due to change of form, we pursue again the method of § 177. From § 76,

$$H_t = \frac{45}{4} \cdot \frac{EI}{ck^2} \cdot tec = \frac{45}{4} \cdot \frac{EI}{ck^2} \cdot \frac{u}{2}.$$

As above,

$$H_t = \frac{12\,EI}{c^2k}v;$$

therefore

$$u = \frac{32}{15} \cdot \frac{k}{c} \cdot v = \frac{8}{15} \cdot \frac{H}{AE} \cdot \frac{3c^2 + 2k^2}{c},$$

a quantity five-sixths of that for the rib with hinged ends.

179. **Circular Rib hinged at Ends.** — It is more difficult to obtain the amount of deflection from change of form produced by the compression at each section of a circular rib, even approximately. As the equilibrium polygon for steady load will not deviate much from the axis of the rib, the thrust T may be assumed to vary as secant θ, the inclination of the rib

at successive points to the horizon: hence the shortening of a small portion, $d\,s$, of arc under the thrust will be

$$d\,(s - s') = \frac{T\,d\,s}{A\,E} = \frac{H\,d\,s}{A\,E}\,\text{secant}\,\theta = \frac{H\,r}{A\,E} \cdot \frac{d\,\theta}{\cos\theta};$$

as the section is constant,

$$s - s' = \frac{H\,r}{A\,E}\int_{-\beta}^{+\beta}\frac{d\,\theta}{\cos\theta} = \log\,\frac{1 + \sin\beta}{1 - \sin\beta} \cdot \frac{H\,r}{A\,E}. \quad (1.)$$

(The symbol *log* denotes the hyperbolic logarithm; to obtain it, multiply the common logarithm by 2.30158.)

As, with a small deflection, the rib will vary but slightly from its original form, let it be assumed to be an arc of a circle after compression. We have then $s - s' = 2\,r\,\beta - 2\,r'\,\beta'$, where r' is the new radius, and β' the new angle subtended by the half-arch. Now

$$r = \frac{c^2 + k^2}{2\,k},\ r' = \frac{c^2 + (k - v)^2}{2\,(k - v)},\ \text{and}\ \sin\beta' = \frac{c}{r'}.$$

By assuming a value for v, r' and β' can be obtained, and the value of $2\,(r\,\beta - r'\,\beta')$ calculated: if it agrees with the value $s - s'$ of equation (1.), the assumed v is sufficiently near the truth; if not, the process of approximation may be repeated. We may adopt, as a value which will answer very well in many cases, $v = \frac{s - s'}{\beta}$. Then

$$v = \frac{H\,r}{A\,E\,\beta}\log\frac{1 + \sin\beta}{1 - \sin\beta}.$$

This logarithmic expression may be written as a series,

$$v = \frac{H\,r}{A\,E\,\beta}\,(\sin\beta + \tfrac{1}{3}\sin^3\beta + \tfrac{1}{5}\sin^5\beta,\ \&c.).$$

It was shown in § 36 that the vertical deflections of two beams of the same cross-section, and carrying the same gross load uniformly distributed, — one inclined at an angle i, and the other the horizontal projection of the former, — were in the pron of 1 : cos i. If, then, the load on the horizontal beam is ..ased in intensity in the ratio sec i : 1, the vertical deflec-ns of the two beams will be the same. We desire to find the ount and distribution of load on a straight beam of the same as the circular arch, Fig. 58, and the same cross-section, h shall produce the same deflection at the middle. By t has just been stated, the load on any horizontal foot of a ght beam must be to the intensity on an inclined beam as c θ to w. A small portion of the arch $d s = \sec \theta \, d x$; hence llows, that, if the arch is carrying w per horizontal foot over whole span, a horizontal beam, as above, loaded with the varying intensity $w \sec \theta = w \frac{d s}{d x}$ per foot, will have the same deflection. This load will be the projection of a load of uniform intensity measured along the rib, or the load on the beam is $w s$, or $2 w r \beta$, in our usual notation.

In any particular case we may easily solve the problem graphically. Lay off 1–2, Fig. 58, equal w . A B; divide A B into a number of equal parts, and 1–2 into the same number, with half-loads at 1 and 2 as usual. Make 2–0 equal to H for this load, and, with 0 as a pole, draw the equilibrium polygon A′ B′, which, for an arch of moderate rise, will be a close approximation to a catenary. C′ B′ . (0–2) will be the desired bending moment M_0, for a deflection found by taking the area moment of A′ B′ C′ about A′, multiplying by 0–2, and dividing by **EI**. Use these values as we did those of § 174. In constructing, increase the length of the rib by (1.) if thought desirable. The values of the following section may be taken if preferred.

180. **Analytical Discussion.** — The exact values may be deduced by the usual process for finding the deflection of a beam. If x is the distance of any point of the beam from one abutment (Fig. 59), β, the angle subtended at the centre by the half-arch, θ, the angle from the crown to any point whose projection is x, and w, the load per foot on the arch, and also at the middle of the beam, then $x = r(\sin \beta - \sin \theta)$, $d x = -r \cos \theta \, d \theta$, the load at any point $= w \sec \theta$ per foot, and load on $d x = w \sec \theta \, d x$

$= - w\,r \sec\theta \cos\theta \, d\theta = - w\,r\,d\theta$. The load on one-half of the span is shown in the figure.

$$\text{Load on half-span} = \int_0^c w \sec\theta \, dx = w\,r \int_0^\beta d\theta = w\,r\,\beta.$$

This expression is the reaction P_1 at the abutment. If x' is the distance from the abutment to any section at which we desire the bending moment, and the corresponding angle is θ', we have the bending moment

$$M = P_1 x' - \int_0^{x'} (x' - x)\, w \sec\theta \, dx$$

$$= w\,r^2 \beta (\sin\beta - \sin\theta') - w\,r^2 \int_\beta^{\theta'} (\sin\theta' - \sin\theta)\, d\theta$$

$$= w\,r^2 (\beta \sin\beta + \cos\beta - \theta' \sin\theta' - \cos\theta'),$$

which becomes at the middle

$$M\,(\max) = w\,r^2 (\beta \sin\beta + \cos\beta - 1) = w\,r\,(c\,\beta - k).$$

Writing the usual expressions for inclination and deflection, and dropping the accents, we have

$$i_x = \int_x^c \frac{M}{EI} dx = -\frac{w\,r^3}{EI} \int_0^\theta (\beta \sin\beta + \cos\beta - \theta \sin\theta - \cos\theta) \cos\theta \, d\theta$$

$$= -\frac{w\,r^3}{EI} (\beta \sin\beta \sin\theta + \cos\beta \sin\theta - \tfrac{3}{4} \sin\theta \cos\theta - \tfrac{3}{4}\theta + \tfrac{1}{2}\theta \cos^2\theta).^*$$

The slope at the abutment, when $\theta = \beta$, is $-\frac{w\,r^3}{4\,EI} (\beta \sin^2\beta - \beta \cos^2\beta + \sin\beta \cos\beta)$,

which, if we remove $\frac{H}{EI}$, is the area of the half equilibrium polygon A′ B′ C′ of Fig. 58. The deflection of the centre is

$$v = \int_0^c i\,dx = \frac{w\,r^4}{EI} \int_0^\beta (\beta \sin\beta \sin\theta + \cos\beta \sin\theta - \tfrac{3}{4} \sin\theta \cos\theta - \tfrac{3}{4}\theta + \tfrac{1}{2}\theta \cos^2\theta) \cos\theta \, d\theta$$

$$= \frac{w\,r^4}{EI} (\tfrac{1}{3} \beta \sin^3\beta + \tfrac{7}{36} \sin^2\beta \cos\beta - \tfrac{1}{4} \beta \sin\beta - \tfrac{1}{9} \cos\beta + \tfrac{1}{9}).^*$$

* These expressions are reduced. To aid any who desire to prove them, we give the following integrals: $\int \theta \cos\theta \, d\theta = \theta \sin\theta + \cos\theta$; $\int \theta \sin\theta \cos\theta \, d\theta = -\tfrac{1}{2} \theta \cos^2\theta + \tfrac{1}{4} \cos\theta \sin\theta + \tfrac{1}{4}\theta$; $\int \cos^2\theta \, d\theta = \tfrac{1}{2} \sin\theta \cos\theta + \tfrac{1}{2}\theta$; $\int \theta \cos^3\theta \, d\theta = \theta \cos^2\theta \sin\theta + \tfrac{7}{9} \cos^3\theta + \tfrac{2}{3} \theta \sin^3\theta + \tfrac{2}{3} \sin^2\theta \cos\theta$.

From this expression, by removing $\frac{H}{EI}$, we obtain the area moment of A′ B′ C′.

The quantities representing v and M will now be introduced in the equation of § 179: hence we get

$$\frac{H}{A\beta}\log\frac{1+\sin\beta}{1-\sin\beta}=\frac{wr^3}{18I}(12\beta\sin^3\beta+7\sin^2\beta\cos\beta-9\beta\sin\beta-4\cos\beta+4).$$

Find the value of M for the special arch, and value of β, and also the value of v. Let $v \div M = B r^2$; then

$$M=\frac{Hr}{2Br^2AE\beta}\log\frac{1+\sin\beta}{1-\sin\beta}.$$

If the arch is a semicircle,

$$M\,(\max)=\tfrac{1}{2}wr^2(\pi-2);\; i=-\frac{wr^3}{4EI}\cdot\frac{\pi}{2};\; v=\frac{wr^4}{36EI}(\tfrac{3}{2}\pi+4).$$

181. **Circular Rib Fixed at Ends.** — From the method of treating the parabolic rib with fixed ends, as compared with the parabolic rib with hinged ends, we would suggest that the deflection and the bending moments at crown and springing of the circular arch with fixed ends, due to the compression of the rib from H, may be obtained from a drawing like Fig. 58, when 2–0 is made equal to the H of this case, by plotting the closing line of Fig. 27 on the arch of Fig. 58, at the height above A of $r\left(\frac{\sin\beta}{\beta}-\cos\beta\right)$ (see § 105), projecting the points of contraflexure vertically on A′ B′, drawing the horizontal closing line of this equilibrium polygon, and then finding M and v for the beam fixed at the ends.

For circular arches of moderate rise, the treatment for parabolic arches will probably suffice.

CHAPTER XII.

BRACED ARCH WITH HORIZONTAL MEMBER; OTHER SPI FORMS; CONCLUSION.

182. **The Usual Analysis not Applicable.** — The difficulty in the way of a successful application of the usual formula $\Sigma \text{E F} . \text{D E} = 0$ for the change of span of the braced arch with horizontal member, of Fig. 60, or, as it is sometimes called, the rib with spandrel bracing, arises from the fact that the *moment of inertia* of successive cross-sections cannot be left out of the equation as a constant. In fact, it varies rapidly; and its amount at any section is unknown until the sizes of the respective pieces are determined. It was shown, in § 72, that **I** must be placed in the denominator of the above formula: and, if not constant, it must come within the sign of summation.

This arch is pivoted at the springings, but continuous at the crown. If it were hinged at the crown by the omission of a piece in either the lower or the upper chord, the thrusts at the abutments could at once be determined by the principles of Chap. II.; and a diagram by the method of Part I., "Roofs," would at once give the stresses in all the pieces for any given load. For the treatment of the case represented in Fig. 60, the following practicable method is offered. It was published in "The Engineer," Feb. 10, 1873, and will also be found in the ninth edition of "The Cyclopaedia Britannica," art. "Bridges," where it is attributed to Professor Clerk-Maxwell.

183. **Change of Span from Stress in a Piece.**—From previous statements, we know that the modulus of elasticity **E** is the measure of the extensibility or compressibility of the kind of material to which it refers, so long as the stress does not surpass the elastic limit, and is equal to the quotient of the intensity of the stress on a cross-section divided by the extension or compression of a *unit's* length of the piece in which the stress is exerted. Thus, if l is the length of a piece in inches, A its cross-section in square inches, T the thrust or tension in pounds to which it is exposed, and Δl the change of length produced,

$$\mathrm{E} = \frac{\mathrm{T}}{\mathrm{A}} \cdot \frac{l}{\Delta l}; \quad \text{or} \quad \Delta l = \frac{\mathrm{T}}{\mathrm{EA}} l. \quad (1.)$$

If the piece A of the frame of Fig. 61 is changed in length, and every other piece is unchanged, while the portion of the frame to the right is held firmly in place, the span L of the frame will undergo an alteration Δ L. In this case the motion takes place about the joint opposite to A, and we may write

$$\Delta \mathrm{L} : \Delta l = ac : ab, \quad (2.)$$

or the distance described by the point b for a small displacement around the axis a will be to the horizontal movement of d as the arm $a\,b$ to the arm of d, or $a\,c$. A similar proportion will be true, if one of the lower chord pieces is supposed to alter in length. In case any diagonal is changed in length, as, for instance, fg, the four-sided figure $efig$ must alter to $efi'g'$ of the sketch below, the point i turning about f as a centre, and the point g about e: hence, for a small displacement, the centre of motion is at the point of meeting, o, of if and ge prolonged, which, for this arch, will lie in the upper chord; and the perpendicular p, dropped on the line of the piece, will take the place of $a\,b$ above.

184. **Stress in a Piece from H and P.**—Let t be the stress produced in a member by a horizontal force H acting between the springing points. Then the principle of equality of moments as necessary for equilibrium about the point around

which motion would otherwise begin, and which is no other than the point noticed at the close of the last section, will determine the relation of the forces. A general rule for finding the axis about which rotation will begin is, Make a section which shall cut three pieces only; prolong the lines of two of the pieces until they meet: the moment of the stress in the third piece about that point of meeting will equal the moment of H about the same point. Hence we have, for the piece A

$$t \,.\, ab = \mathrm{H} \,.\, a\,c, \text{ or } t = \frac{a\,c}{a\,b}\mathrm{H}.$$

Similarly, let t' be the stress produced in A by a vertical force P applied at one springing, while the other end of the frame is held rigidly so that it cannot turn. As the arm of P will be $d\,c$, we may write

$$t' \,.\, ab = \mathrm{P} \,.\, d\,c, \text{ or } t' = \frac{d\,c}{a\,b}\mathrm{P}.$$

The distances $d\,c$ and $a\,c$, being respectively horizontal and vertical, may be denoted in general for any piece by x and y. In order to make the symbol $a\,b$ of the last section and of this one general, so as to apply to a diagonal as well as a chord piece, let us write for $a\,b$ the perpendicular p drawn from the axis of rotation upon the action-line of the piece.

Any thrust at the springing having horizontal and vertical components H and P will produce a stress T in the piece, equal to $t + t'$, or

$$\mathrm{T} = \frac{a\,c \,.\, \mathrm{H} + d\,c \,.\, \mathrm{P}}{a\,b} = \frac{\mathrm{H}y + \mathrm{P}x}{p}. \quad (1.)$$

It is evident that heed must be paid to the kind of stress produced by H and P; thus, in any piece of the top member, H will produce tension and elongation, while P will produce compression and shortening: the reverse will be true of the lower member; how the diagonals are affected will be seen when we come to our application. Appropriate signs, therefore, must be given to the arithmetical values of the stress and alter-

of length; thus compression and shortening may be positive; tension and lengthening, negative.

Formula for H. — From equations (1.) and (2.), § 183, writing y and p, as indicated above, for $a\,c$ and $a\,b$, we get change of span for any stress, T, in a particular piece,

$$\Delta L = \Delta l \frac{y}{p} = \frac{T y}{p} \cdot \frac{l}{E A},$$

upon inserting the value of T from equation (1.), last sec ,

$$\Delta L = \frac{H y^2 + P x y}{p^2} \cdot \frac{l}{E A}.$$

This same quantity can be calculated for the extensibility due to each member of the frame; and the result will not be altered by the slight yielding of all the others, unless this yielding produces sensible deformation, making appreciable changes in $\frac{x}{p}$ and $\frac{y}{p}$: hence the sum of all the changes of span, or the total change of span, will be

$$H \Sigma \frac{y^2}{p^2} \cdot \frac{l}{E A} + \Sigma P \frac{x y}{p^2} \cdot \frac{l}{E A}.$$

If the abutments do not yield, this expression is zero. If the span changes, by a yielding of the abutments, so that e is the elongation of span for one ton of H, then the above expression for change of span equals e H. P is the vertical component of the reaction at one abutment, found as for any frame loaded as this arch may be: hence H may be found. If the abutments do not yield, we then obtain

$$H = \frac{\Sigma P \frac{x y}{p^2} \cdot \frac{l}{E A}}{\Sigma \frac{y^2}{p^2} \cdot \frac{l}{E A}}. \quad (1.)$$

186. **Application of Method.** — Let a single weight, W, be applied at any one of the top joints of the braced arch, Fig. 60.

Inclined reactions will be produced at each abutment, whose components will be H and P_1 at the left, H and P_2 at the right. The calculations for the resulting stresses in the pieces are then best made as follows: Construct tables of the values $x \div p$ and $y \div p$ for each member of the frame; the method of sections through the opposite joints, or of moments, will answer best for the top and bottom members, and a diagram such as has been drawn for a roof, for the diagonals; assume a cross-section for each member for an assumed probable value of the abutment thrust; make tables of $\frac{xy}{p^2} \cdot \frac{l}{\text{E A}}$ and $\frac{y^2}{p^2} \cdot \frac{l}{\text{E A}}$, or, what is equivalent when all the frame is of one material, so that **E** is constant, make tables of $\frac{x\,y\,l}{p^2\,\text{A}}$ and $\frac{y^2\,l}{p^2\,\text{A}}$. The summations indicated in (1.), § 185, can then be made. In summing $\text{P} \cdot \frac{x\,y\,l}{p^2\,\text{A}}$, the value P_1 must be used for all pieces to the left of the loaded joint, and P_2 for all pieces to the right of the load. Equation (1.), above, will now give the value of H for this single load.

The process of finding the numerator of (1.) must be repeated for each joint which is loaded. The abutment reactions having thus been found, the stress in each piece will be computed by (1.) § 184, or will be scaled from a diagram drawn as in Part I., "Roofs." If, upon finding the maximum stresses in the pieces, resulting from the steady load and such rolling loads as will have the worst effect, the assumed sections are not strong enough for these stresses, fresh cross-sections must be assumed, and the whole calculation repeated. The change in cross-sections will cause some change in the values of H; but this tentative process need seldom be repeated but once.

187. **Example; Stresses from H and P.** — These processes will probably be rendered more clear by an example. Let the arched frame of Fig. 60 be 120 feet in span, 12 feet rise to the curved member, and 17 feet rise to the straight member, making the depth at mid-span 5 feet. Let the upper member be divided into panels of 10 feet each, and the parabolic or circu-

lar arc into portions of 10.263 feet each.[1] The radius of the curved member will be 156 feet. Let it be desired to design this arched structure to bear a steady load of ten tons per joint of the top member and a travelling load of the same intensity.

If a horizontal line L O is drawn to represent a certain value of H, we may construct Fig. 62 by the method used in Part I., "Roofs," and by scale determine the magnitude of the stress in each piece due to this H, as the *only force*, applied as a thrust at each abutment; all of the stresses being measured as *fractions of* H, and the kind of stress noted. One-half of the diagram is sufficient, as it will be symmetrical. The magnitude of any stress in a top or bottom piece can be readily proved by the method of moments. We may now fill the columns of a table with these ratios which represent $y \div p$, being not only the ratios of the stresses to H, but of the change of span to change of length. Bow's notation is used, and the stresses in one half of the frame will correspond with those in the other half. The sign + denotes compression, the sign − denotes tension.

VALUES OF $\frac{y}{p}$.

B O −0.272	A L +1.203	O A −0.444	A B +0.450
D O −0.639	C L +1.520	B C −0.478	C D +0.480
F O −1.117	E L +1.927	D E −0.500	E F +0.502
I O −1.678	G L +2.427	F G −0.484	G I +0.488
K O −2.185	J L +2.942	I J −0.384	J K +0.386
N O −2.400	M L +3.293	K M −0.153	M N +0.154

In the same way a diagram constructed upon a vertical line which represents P_1, Fig. 63, will give the stresses in the several pieces caused by this vertical force only, applied in an upward direction at the left abutment, while the right end is held rigidly in place by fixing the end brace in position. This figure will not be symmetrical, and therefore all the pieces must be entered in the table. P_2 at the right abutment, in place of P_1 at the left, will reverse the table, B′ O taking the place of B O, &c. The ratio of these stresses to P will give $x \div p$.

[1] If the arc is parabolic, the length of a piece will be 10.268 feet. The difference is not material for our example.

VALUES OF $\frac{x}{p}$.

B O	+ 0.718	A L	− 0.354	O A	+1.178	A B	−1.189
D O	+ 1.872	C L	− 1.341	B C	+1.505	C D	−1.780
F O	+ 3.662	E L	− 2.833	D E	+1.872	E F	−1.879
I O	+ 6.226	G L	− 4.996	F G	+2.214	G I	−2.282
K O	+ 9.319	J L	− 7.787	I J	+2.341	J K	−2.353
N O	+12.000	M L	−10.655	K M	+1.907	M N	−1.920
K′ O	+13.163	M′ L	−12.592	N M′	+0.833	M′ K′	−0.827
I′ O	+12.675	J′ L	−12.978	K′ J′	−0.371	J′ I′	+0.309
F′ O	+11.283	G′ L	−12.134	I′ G′	−1.212	G′ F′	+1.202
D′ O	+ 9.698	E′ L	−10.767	F′ E′	−1.657	E′ D′	+1.664
B′ O	+ 8.260	C′ L	− 9.387	D′ C′	−1.876	C′ B′	+1.880
		A′ L	− 8.139	B′ A′	−1.367	A′ O	fixed.

188. **Computation of Tables.** — We may now write a table for $\frac{y^2 l}{p^2}$, and another for $\frac{x\,y\,l}{p^2}$, for each piece of the frame. The first table, involving squares, will be positive throughout. The lengths of the horizontal and rib pieces will be multiplied by the footing of their respective columns to save labor; but the lengths of the diagonals are carried in as indicated.

VALUES OF $\frac{y^2 l}{p^2}$.

B O	0.074	A L	1.447	O A	0.197	× 17.72	= 3.491	A B	0.202	× 14.08	= 2.844
D O	0.408	C L	2.310	B C	0.228	× 14.33	3.267	C D	0.230	× 11.17	2.569
F O	1.248	E L	3.713	D E	0.250	× 11.58	2.895	E F	0.252	× 9.15	2.306
I O	2.816	G L	5.890	F G	0.234	× 9.67	2.263	G I	0.238	× 7.75	1.844
K O	4.774	J L	8.655	I J	0.147	× 8.25	1.213	J K	0.140	× 7.17	1.068
N O	5.760	M L	10.844	K M	0.023	× 7.50	0.172	M N	0.024	× 7.07	0.170
	15.080 × 10		32.859				13.301				10.801
	9.320 × 10		2				2				2
	244.000		65.718 × 10.263 = 674.46				26.602				21.602

Summing these columns, and doubling for the whole arch, we obtain $244.00 + 674.46 + 26.60 + 21.60 = 966.66 = \Sigma . \frac{y^2 l}{p^2}$. If, in the first trial, all the sections are supposed equal, A may be omitted from (1.), § 185, and 966.66 becomes the denominator of that expression.

We next compute the following table, and multiply by the length of each piece as we advance. It will be convenient to add other columns, marked Σ, containing successive summations of the factors for each set of pieces, as these numbers will be used in turn. The *summations* are all *negative*, as will be readily seen, and hence the sign — is omitted.

VALUES OF $\frac{x y l}{p^2}$.

	Σ		Σ		Σ		Σ
B O — 1.95	1.95	A L — 4.37	4.37	O A — 9.27	9.27	A B — 7.53	7.53
D O — 11.96	13.91	C L — 20.92	25.29	B C — 10.30	19.57	C D — 9.54	17.07
F O — 40.90	54.81	E L — 56.05	81.34	D E — 10.84	30.41	E F — 8.63	25.70
I O — 104.47	159.28	G L — 124.44	205.78	F G — 10.37	40.78	G I — 8.44	34.14
K O — 203.62	362.90	J L — 235.11	440.89	I J — 7.42	48.20	J K — 6.51	40.65
N O — 288.00	650.90	M L — 360.10	800.99	K M — 2.19	50.39	M N — 2.09	42.74
K′O — 287.61	938.51	M′ L — 425.55	1226.54	N M′ + 0.95	49.44	M′K′ + 0.10	42.64
I′ O — 212.69	1151.20	J′ L — 391.85	1618.39	K′ J′ — 1.17	50.61	J′ I′ — 0 85	43.49
F′O — 126.03	1277.23	G′ L — 302.23	1920.62	I′ G′ — 5.68	56.29	G′ F′ — 4.55	48.04
D′O — 61.97	1339.20	E′ L — 212.94	2133.56	F′ E′ — 9.59	65.88	E′ D′ — 7.64	55 68
B′O — 22.47	1361.67	C′ L — 146.43	2280.00	D′ C′ — 12.85	78.73	C′ B′ — 10 07	65.75
		A′ L — 100.48	2380.48	B′ A′ — 10.76	89.49	A′ O fixed.	——

189. **Values of H.** — The calculations for H can now be proceeded with, and they are given below. An explanation of one computation will suffice for all. If a weight W is placed on the third upper joint from the left, the vertical component of the left abutment reaction, P_1, is $\frac{19}{24}$ W. Then, for the two pieces of the upper chord to the left we have $\Sigma P_1 \frac{x y}{p^2} l = 13.91 P_1$; for the two pieces of the rib to the left, we get 25.29 P_1, and, for the five web-members to the left, $30.41 + 17.07 = 47.48 P_1$. On the right of the weight, the nine remaining pieces of the upper chord give $\Sigma P_2 \frac{x y}{p^2} l = 1277.23 P_2$, which will be found opposite F′ O, as the vertical force is now applied at the right end; for the ten pieces of the rib we find 2133.56 P_2, and for the rest of the web to E F we find opposite E′ F′ and F′ G′, for the reason

just stated, $65.88 + 48.04 = 113.92\ P_2$. As the piece E L, below the weight, is acted upon by P_1 on one side, and P_2 on the other, it makes no difference whether it is considered to lie to the left or the right of the loaded point. Adding up the respective numbers, multiplying one by $\frac{19}{24}$, and the other by $\frac{5}{24}$, adding, and dividing by $\Sigma \frac{y^2}{p^2} l = 966.66$, we get $H = 0.831\ W$ for a load on the third joint only. The divisor $966.66 \times 24 = 23{,}200$, is used.

VALUES OF H.

W on 1st Joint.		W on 2d Joint.		V	
0	1361.67	1.05	1339.20	13.	
0	2380.48	4.37	2280.00	25.29	
9.27	80.49	19.57	78.73	30.41	[illegible]
9.27	65.75	7.53	55.68	17.07	48.[illegible]
23	3897.39	33.42	3753.61	86.68	3524.71
213.21		21	3	19	5
3897.39		701.82	11260.83	1646.92	17623.55
41.1060÷232=.177 W		11260.83		17623 55	
		119.6265÷232=.516 W.		192.7047÷232=831 W.	

W on 4th Joint.		W on 5th Joint.		W on 6th Joint.	
54.81	1151.20	159.28	938.51	362.90	650.90
81.34	1920.62	205.78	1618.39	440.89	1226.54
40.78	56.29	48.20	50.61	50.39	49.44
25.70	43.49	34.14	42.04	40.65	42.74
202.63	3171.60	447.40	2650.15	894.83	1969.62
17	7	15	9	13	11
3444.71	22201.20	6711.00	23851.35	11632.79	21665.82
22201.20		23851.35		21665.82	
256.4591÷232=1.105 W.		305.6235÷232=1.317 W.		333.0861÷232=1.436W.	

Having completed the computations for six joints, we add the H's, and multiply by two, obtaining 10.764 W as the value of H for an entire load of W on each upper joint.

We next compute the following table, and multiply by the length of each piece as we advance. It will be convenient to add other columns, marked Σ, containing successive summations of the factors for each set of pieces, as these numbers will be used in turn. The *summations* are all *negative*, as will be readily seen, and hence the sign — is omitted.

VALUES OF $\frac{x\,y\,l}{p^2}$.

	Σ		Σ		Σ		Σ
1.95	1.95	A L — 4.37	4.37	O A — 9.27	9.27	A B — 7.53	7.53
11.96	13.91	C L — 20.92	25.29	B C —10.30	19.57	C D — 9.54	17.07
— 40.90	54.81	E L — 56.05	81.34	D E —10.84	30.41	E F — 8.63	25.70
O —104.47	159.28	G L —124.44	205.78	F G —10.37	40.78	G I — 8.44	34.14
K O —203.62	362.90	J L —235.11	440.89	I J — 7.42	48.20	J K — 6.51	40.65
N O —288.00	650.90	M L —360.10	800.99	K M — 2.19	50.39	M N — 2.09	42.74
K′O —287.61	938.51	M′L —425.56	1226.54	N M′+ 0.05	49.44	M′K′ + 0.10	42.64
I′ O —212.69	1151.20	J′ L —391.85	1618.39	K′ J′— 1.17	50.61	J′ I′ — 0 85	43.49
F′O —126.03	1277.23	G′ L —302.23	1920.62	I′ G′— 5.68	56.29	G′ F′ — 4.55	48.04
D′O — 61.97	1339.20	E′ L —212.94	2133.56	F′E′— 9.59	65.88	E′D′ — 7.64	55.68
B′O — 22.47	1361.67	C′ L —146.43	2280.00	D′ C′—12.85	78.73	C′ B′ —10 07	65.75
		A′ L —100.48	2380.48	B′ A′—10.76	89.49	A′ O fixed.	—

189. **Values of H.** — The calculations for H can now be proceeded with, and they are given below. An explanation of one computation will suffice for all. If a weight W is placed on the third upper joint from the left, the vertical component of the left abutment reaction, P_1, is $\frac{19}{24}$ W. Then, for the two pieces of the upper chord to the left we have $\Sigma P_1 \frac{x\,y}{p^2} l = 13.91\ P_1$; for the two pieces of the rib to the left, we get 25.29 P_1, and, for the five web-members to the left, $30.41 + 17.07 = 47.48\ P_1$. On the right of the weight, the nine remaining pieces of the upper chord give $\Sigma P_2 \frac{x\,y}{p^2} l = 1277.23\ P_2$, which will be found opposite F′ O, as the vertical force is now applied at the right end; for the ten pieces of the rib we find 2133.56 P_2, and for the rest of the web to E F we find opposite E′ F′ and F′ G′, for the reason

just stated, 65.88 + 48.04 = 113.92 P_2. As the piece E L, below the weight, is acted upon by P_1 on one side, and P_2 on the other, it makes no difference whether it is considered to lie to the left or the right of the loaded point. Adding up the respective numbers, multiplying one by $\frac{19}{24}$, and the other by $\frac{5}{24}$, adding, and dividing by $\Sigma \frac{y^2}{p^2} l = 966.66$, we get H = 0.831 W for a load on the third joint only. The divisor 966.66 × 24 = 23,200, is used.

VALUES OF H.

W on 1st Joint.		W on 2d Joint.		W on 3d Joint.	
0	1361.67	1.95	1339.20	13.91	1277.28
0	2380.48	4.37	2280.00	25.29	2133.56
9.27	80.49	10.57	78.73	30.41	65.88
9.27	65.75	7.53	55.68	17.07	48.04
23	3897.39	33.42	3753.61	86.68	3524.71
213.21		21	3	19	5
3897.39		701.82	11260.83	1646.92	17623.55
41.1060+232=.177 W		11260.83		17623.55	
		119.6265+232=.516 W.		192.7047+232=831 W.	

W on 4th Joint.		W on 5th Joint		W on 6th Joint.	
54.81	1151.20	159.28	938.51	362.90	650.90
81.34	1920.62	205.78	1618.39	440.89	1226.54
40.78	56.29	48.20	50.61	50.39	49.44
25.70	43 49	31.14	42.64	40.65	42.74
202.63	3171.60	417.40	2650.15	894.83	1969.62
17	7	15	9	13	11
3444.71	22201.20	6711.00	23851.35	11632.79	21665.82
22201.20		23851.35		21665.82	
256.4591+232=1.105 W.		305.6235+232=1.317 W.		333.0861+232=1.436W.	

Having completed the computations for six joints, we add the H's, and multiply by two, obtaining 10.764 W as the value of H for an entire load of W on each upper joint.

Stresses in Pieces, All Cross-Sections Equal.

Load on	BO.	DO.	FO.	IO.	KO.	NO	AL.	CL.	EL.	GL.	JL.	ML.
1st.	+0.30	+0.30	+0.28	+0.24	+0.17	+0.09	−0.13	−0.13	−0.10	−0.07	−0.01	+0.07
2d.	+0.49	+0.89	+0.85	+0.74	+0.54	+0.29	+0.30	−0.39	−0.35	−0.25	−0.08	+0.14
3d.	+0.34	+0.96	+1.43	+1.26	+0.95	+0.53	+0.70	+0.17	−0.69	−0.57	−0.31	+0.06
4th.	+0.21	+0.62	+1.36	+1.85	+1.43	+0.86	+1.07	+0.72	+0.11	−0.89	−0.56	−0.04
5th.	+0.09	+0.33	+0.82	+1.68	+2.06	+1.34	+1.37	+1.15	+0.73	+0.02	−1.07	−0.44
6th.	0.00	+0.10	+0.38	+0.97	+1.91	+2.05	+1.53	+1.43	+1.20	+0.73	−0.07	−1.16
7th.	−0.06	−0.06	+0.07	+0.44	+1.13	+2.05	+1.54	+1.52	+1.40	+1.12	+0.56	−0.26
8th.	−0.09	−0.14	−0.10	+0.12	+0.61	+1.34	+1.45	+1.47	+1.45	+1.30	+0.93	+0.32
9th.	−0.09	−0.16	−0.16	−0.03	+0.31	+0.86	+1.22	+1.27	+1.29	+1.21	+0.98	+0.52
10th.	−0.07	−0.13	−0.16	−0.08	+0.14	+0.53	+0.90	+0.95	+0.97	+0.93	+0.78	+0.46
11th	−0.05	−0.09	−0.10	−0.07	+0.06	+0.29	+0.55	+0.60	+0.62	+0.62	+0.53	+0.37
12th.	−0.02	−0.03	−0.04	−0.03	0.00	+0.09	+0.19	+0.20	+0.22	+0.21	+0.19	+0.15
Σ +	1.43	3.20	5.10	7.30	9.31	10.32	10.82	9.48	7.99	6.14	3.97	2.09
Σ −	0.38	0.61	0.56	0.21	0.00	0.00	0.13	0.52	1.14	1.78	2.10	1.90
S. L.	+1.05	+2.59	+4.03	+7.09	+9.31	+10.32	+10.69	+8.96	+6.85	+4.36	+1.87	+0.19
Max.	+2.48	+5.79	+9.82	+14.39	+18.02	+20.64	+21.51	+18.44	+14.84	+10.50	+5.84	+2.28
											−0.23	−1.71

Load on	OA.	AB.	BC.	CD	DE	EF.	FG.	GI.	IJ.	JK.	KM.	MN.
1st.	+1.05	+0.01	−0.01	+0.01	−0.03	+0.03	−0.04	+0.04	−0.06	+0.06	−0.06	+0.06
2d.	+0.80	−0.82	+1.07	+0.02	−0.07	+0.07	−0.09	+0.00	−0.16	+0.16	−0.18	+0.19
3d.	+0.58	−0.58	+0.81	−0.82	+1.07	+0.07	−0.14	+0.15	−0.25	+0.25	−0.29	+0.30
4th	+0.33	−0.34	+0.53	−0.54	+0.77	−0.79	+1.04	+0.17	−0.32	+0.33	−0.42	+0.42
5th.	+0.16	−0.16	+0.32	−0.33	+0.53	−0.51	+0.76	−0.78	+0.95	+0.39	−0.52	+0.52
6th	0.00	0.00	+0.11	−0.15	+0.31	−0.32	+0.51	−0.52	+0.72	−0.74	+0.84	+0.60
7th	−0.06	+0.06	+0.03	−0.04	+0.15	−0.16	+0.33	−0.31	+0.52	−0.55	+0.65	−0.66
8th	−0.13	+0.13	−0.01	+0.04	+0.04	−0.04	+0.18	−0.19	+0.37	−0.38	+0.51	−0.53
9th.	−0.13	+0.14	−0.07	+0.07	−0.02	+0.02	+0.11	−0.12	+0.25	−0.26	+0.38	−0.39
10th.	−0.10	+0.10	−0.07	+0.07	−0.02	+0.02	+0.07	−0.07	+0.17	−0.18	+0.27	−0.28
11th.	−0.07	+0.08	−0.05	+0.05	−0.03	+0.03	+0.02	−0.02	+0.09	−0.09	+0.15	−0.15
12th.	−0.02	+0.02	−0.01	+0.01	0.00	0.00	+0.01	−0.01	+0.02	−0.02	+0.05	−0.05
Σ +	2.92	0.54	2.90	0.27	2.87	0.24	3.03	0.45	3.09	1.19	2.85	2.00
Σ −	0.51	1.90	0.25	1.87	0.17	1.85	0.27	2.05	0.79	2.22	1.47	2.06
S. L.	+2.41	−1.36	+2.65	−1.60	+2.70	−1.61	+2.76	−1.60	+2.30	−1.03	+1.38	0.00
Max.	+5.33		+5.55		+5.57		+5.79		+5.30	+0.16	+4.23	+2.09
		−3.26		−3.47		−3.46		−3.65		−3.25	−0.09	−2.06

VALUES OF $\frac{y^2}{p^3} \cdot \frac{l}{A}$.

B O	0.296	A L	0.069	O A	0.582	A B	0.948	
D O	0.680	C L	0.128	B C	0.544	O D	0.642	15.971
F O	1.248	E L	0.248	D E	0:483	E F	0.577	134.527
I O	1.877	G L	0.535	F G	0.377	G I	0.461	4.544
K O	2.513	J L	1.236	I J	0.243	J K	0.356	6.138
N O	2.743	M L	4.338	K M	0.043	M N	0.085	161.180
	9.357		6.554		2.272		3.069	24
	6.614		2		2		2	3868.320
	15.971		13.108 × 10.263		4.544		6.138	

VALUES OF $\frac{xy}{p^3} \cdot \frac{l}{A}$.

		Σ			Σ			Σ			Σ
B O	− 0.78	0.78	A L	− 0.21	0.21	O A	−1.54	1.54	A B	−2.51	2.51
D O	− 1.99	2.77	C L	− 1.16	1.37	B C	−1.72	3.26	C D	−2.38	4.89
F O	− 4.09	6.86	E L	− 3.74	5.11	D E	−1.81	5.07	E F	−2.16	7.05
I O	− 6.96	13.82	G L	− 11.31	16.42	F G	−1.73	6.80	G I	−2.11	9.16
K O	−10.72	24.54	J L	− 33.59	50.01	I J	−1.48	8.28	J K	−2.17	11.33
N O	−13.71	38.25	M L	−144.04	194.05	K M	−0.55	8 83	M N	−1.05	12.38
K′O	−15.14	53.39	M′L	−170.22	364.27	N M′	+0.48	8.35	M′K′	+0 02	12.36
I′ O	−14.18	67.57	J′ L	− 55.98	420.25	K′J′	−0.39	8.74	J′ I′	−0 17	12 53
F′O	−12.60	80.17	G′L	− 27.48	447.73	I′ G′	−1.42	10.16	G′ F′	−0 73	13 29
D′O	−10.33	90.50	E′ L	− 14.20	461.93	F′E′	−2.40	12 56	E′ D′	−1.27	14.56
B′O	− 8.99	99.49	C′ L	− 8.13	470.06	D′C′	−3.21	15.77	C′ B′	−1 68	16.24
			A′L	− 4.78	474.84	B′ A′	−3.58	19.35	A′O′ fixed.		

The above *summations* are negative.

Next follow, as before, the computations of H (p. 185).

It will be seen that the change in the sections of the pieces has made but little change in the values of H; the thrust now being 10.820 W for a steady load of W on each joint. We may therefore proceed to draw anew the diagrams for a single load W on any one joint, or we may, by the use of lines of another color, alter the figures already drawn. As H has been changed so little, the new stresses will determine the final

Values of H.

W on 1st Joint.		W on 2d Joint.		W on 3d Joint.	
0.00	99.49	0.78	90.50	2.77	80.17
0.00	474.84	0.21	470.06	1.37	461.03
1.54	19.85	3.26	15.77	5.07	12.56
1.54	16.24	2.51	14.56	4.89	13.29
23	609.92	6.76	590.89	14.10	567.95
35.42		21	8	19	5
609.92		141.96	1772.67	267.90	2839.75
645.34+3868=.167 W.		1772.67		2839.75	
		1914.63+3868=.495 W.		3107.65+386[illegible]	

W on 4th Joint.		W on 5th Joint.		W on 6th Joint.	
6.86	67.57	13.82	53.39	24.54	38.25
5.11	447.73	16.42	420.25	50.01	364.27
6.80	10.16	8.28	8.74	8.83	8.35
7.05	12.53	9.16	12.36	11.33	12.38
25.82	537.99	47.68	494.74	94.71	423.25
17	7	15	9	13	11
438.94	3765.93	715.20	4452.66	1231.23	4655.75
3765.93		4452.66		4655.75	
4204.87+3868=1.087 W.		5167.86+3868=1.336 W.		5886.98+3868=1.522 W.	

dimensions of the pieces. A sample of the stresses obtained in the upper chord is given below for comparison.

	B O.	D O.	F O.	I O.	K O.	N O.
Σ +	1.45	3.18	5.10	7.08	9.23	10.20
Σ −	0.42	0.63	0.51	0.07	0.00	0.00
S. L.	1.03	2.55	4.59	7.01	9.23	10.20
Max.	+2.48	+5.73	+9.69	+14.09	+18.46	+20.40

A certain allowance in section may be made for the stresses from change of temperature, or the effect of the change of length in each piece may be worked out separately.

192. **Bracing with Vertical Struts.** — The bracing of the arch just described is of the Warren or triangular type. The design of Fig. 65 has been used with success, is probably more economical of material, and is, in our judgment, more pleasing to the eye. The inclined braces are ties, and the introduction of the counters at the crown obviates the reversal of stress in the braces. When the upper member approaches the curved member closely at the crown, the web may be made of a plate for a distance of two panels: sometimes the two members are brought into contact at the crown.

193. **Cast-Iron Arch as a Breast-Summer.** — Builders sometimes employ a cast-iron member, shaped like Fig. 66, for spanning openings of considerable size, and carrying the weight of a brick wall. Aside from the fact that cast-iron in large masses of very uncertain strength, by reason of internal stresses produced by contraction in cooling, an additional element of uncertainty is introduced by the method of constructing these ribs. The thrust of the arch is resisted by a wrought-iron rod, represented by a straight line in the figure, which, in place of being fastened by bolts or nuts, is fitted into recesses in the casting at its ends. In order to have the rod tight, it is made shorter than the distance between bearings, then heated, and shrunk into place. The rod is therefore under an initial tension, and the rib under initial compression, both of which are likely to be of uncertain amount, and detrimental; for, when the arch is loaded, its horizontal thrust will be added to the tension in the bar, and the compression of the rib will be increased. As, however, the bar elongates under the pull, it would be well, were it possible, to have the bar so much shorter than the normal span of the arch, that the value of H proper to the arch under the proposed load should elongate the rod to that normal span; then the initial bending moments produced in the rib by shrinking on the rod will be removed. It would seem possible, by a careful measurement of the extension of the rod between two marks some ten or twenty feet apart, especially if the stretch has been previously tested, to determine the initial tension.

If the arch is well built into the masonry at the ends, and if the bearings are long, the rib may be considered as fixed at the ends. If not so built, and in preliminary testing on two supports under an applied weight, the rib must be considered as pivoted at the ends. From the small rise, such arches may be assumed, in either case, to be parabolic. In testing, therefore, under a single weight W applied at the middle, by § 40 $H = \frac{25}{64}\frac{c}{k}W$. At that time temporary bearings ought to be placed at A to prevent the arch from bearing at C when loaded. Under the load of the wall, unless the latter is cut by large openings, so that a pier concentrates the weight on a small portion of the rib, there will be no bending moments, as the load is uniformly distributed.

194. **Gothic Rib for Roofs.**—The rib which supports the roof of the Grand Central Depot in New-York City is probably circular, and will be analyzed readily by the principles already laid down; but the Gothic rib requires some special treatment. Fig. 67 is a sketch of the rib which sustains the roof over the train-house of the Boston and Providence Railroad Depot in Boston, Mass. The span is 125 feet between walls, and the height is 55 feet to the axis of the rib. As height impresses one more than horizontal distance, it is evident that this roof appears lofty when viewed from the inside. In order to give height quickly near the walls, the half-rib is struck with two radii, as indicated in the figure. The lower portion is built with a solid web; while most of the upper portion has a uniform depth of three feet. If the junction at the crown or apex of the roof allows any movement, if the ribs can rock or turn on castings at their bases, and if they are independent of the side walls, they may be treated as hinged at three points, and discussed like any three-hinged arch. If there is no opportunity for movement at the bases, and especially if the ribs abut closely against the side walls and buttresses, while still a joint is provided at the crown, the condition of invariability of span must be applied, and also the condition that the deflection of

the crown when measured by area moments from the tangent at one abutment shall equal the deflection of the crown from the tangent at the other abutment. The integration will then be between limits which will appear from the discussion of the third supposition.

The rib may be fixed at the ends and crown, and will then offer a troublesome case for treatment by reason of the great depth at the haunches, unless we assume that it is well buttressed by the wall. In this case, the portion below the top of the wall and the wall itself will act as an abutment; and, as it will only require a moderate tension in the inside flange at the springing to resist the overturning moment, such an assumption seems entirely warrantable. Above the wall, then, some 25 feet high, where the horizontal mark is made on the left-hand side, we assume the springing line of the arch, and consider the remainder as a rib fixed at the ends, and continuous at the crown. In applying the conditions for a rib with fixed ends to this case, we must change the derived equations, as the curve is not continuous at the crown. A parabola drawn through the middle of the depth of the rib at crown, springing, and a third point near the upper end of the straight portion of the rafter, will agree very closely with the axis of the rib throughout. We must first determine k and c for this parabola. In Fig. 68 let h be the height or rise of the arch at the apex, a the horizontal distance from h to the point where the parabola would become horizontal; then

$$h = \frac{k}{c^2}(c^2 - a^2); \quad \text{or} \quad k = h\frac{c^2}{c^2 - a^2}.$$

For another ordinate h', distant $c - a'$ from the springing, we write

$$k = h'\frac{c^2}{c^2 - a'^2}.$$

In this case $c - a = 55.75$ feet, $h = 30.3$ feet, $c - a' = 22.5$ feet, and $h' = 17$ feet: hence we find that $k = 31.68$ feet, $c = 70.48$ feet, and $a = 14.73$ feet.

In place of performing the integrations of §§ 58–59 between the limits there given, we must omit or subtract from the equations the integrals between the limits $+a$ and $-a$, as this portion is cut out of the parabola. Thus the equation (1.) of § 58 will be written

$$\int_0^{2c} \text{D E}^2 - \int_{c-a}^{c+a} \text{D E}^2 = \int_0^{c+b} \text{D F . D E} - \int_{c-a}^{c+a} \text{D F . D E} + \int_0^{c-b} \text{D F . D E.}$$

As limits $c+a$ and $c-a$ will yield terms similar to limits $c+b$ and $c-b$, the subtractive quantities above can be written from inspection of (2.), § 58, and (2.), § 39. A similar treatment of the other equations of condition will be required. The solution will then proceed as usual.

If the weight at the apex of the roof, arising from the ventilator, &c., is sufficiently great, it will take the place of the omitted portion of breadth $2a$, so that the rib will be very nearly in equilibrium under steady load.

195. **Remarks on Designing.** — The examples which have been given in the preceding pages will indicate the steps to be pursued in working out a specific design. The type of structure having been determined upon, the moving load must be taken of an intensity in harmony with the position of the bridge, or we must decide upon the weight of snow and pressure of wind to which the roof will be liable. The dead weight of the structure must then be assumed, of such an amount as our judgment and experience dictate, to be afterwards verified and corrected from the actual sections. The abutment reactions and bending moments from the applied forces will then be found, after which, stress diagrams may be constructed, or equilibrium polygons drawn; from the first we obtain stresses directly, as in Part I.; from the second, bending moments, with shears and direct thrusts, from which the stresses in the several pieces will be found, as in Part II. The first method is probably the shorter for roofs, unless the rib is solid, or has a plate web, as all of the load of one kind may be included at one operation: the second method will be preferred where a moving load has to be

considered. The stresses will then be tabulated, and the maximum compression and tension on each piece found.

A point which may call for a little explanation is illustrated by Fig. 69. We desire to draw a stress diagram for an arched rib, which is fixed at the end A B, the equilibrium curve beginning with the line G D, and the bending moment at A B being T . p, or its equivalent. The flanges at A and B will transmit direct force only: therefore decompose T into C, the compression parallel to the flanges, at the springing, and F, the shear at right angles. Then, by moments about A, Thrust at B . A B = C . A G, or Thrust at $B = \frac{C \cdot AG}{AB}$; by moments B, Tension at $A = \frac{C \cdot BG}{AB}$. The shear F will be re[illegible]ted either at A or B, depending upon which of the braces is designed to carry it: if the braces are ties, it must pass through the one at A. Thus we obtain the forces with which to begin the stress diagram. In case of a hinge at the abutment, the point G is found midway between A and B, and there will be $\frac{1}{2}$ C, compression, at each flange. F will be found in the proper brace as above.

The arched rib must be thoroughly stayed laterally; for so much of either flange as is compressed is in unstable equilibrium; between lateral stays, the breadth of a compressed flange must be determined from the formulæ for columns. For a few formulæ and directions for detailing, see the closing chapter of Part I.

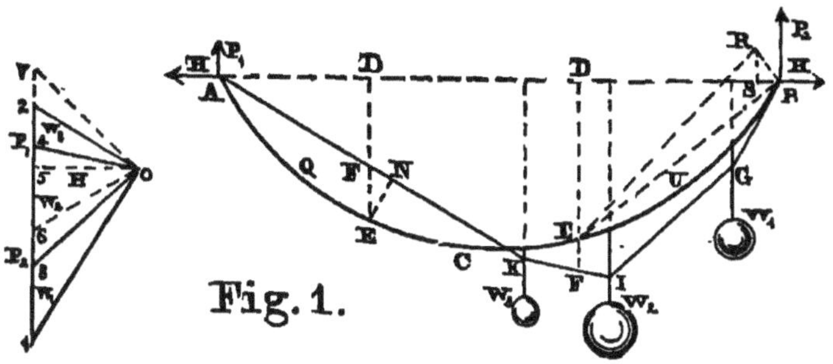

Fig. 1.

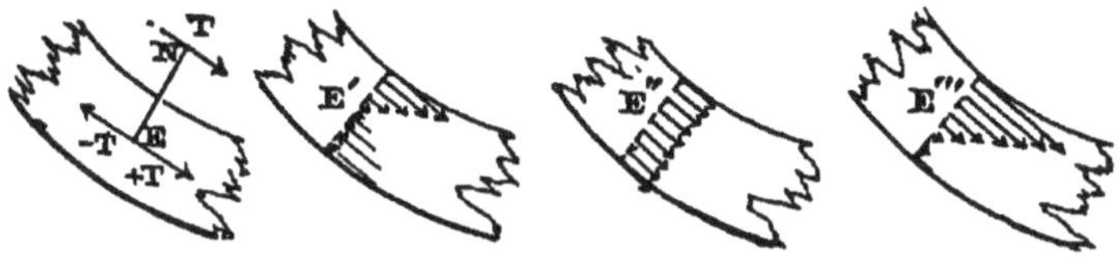

Fig. 2.

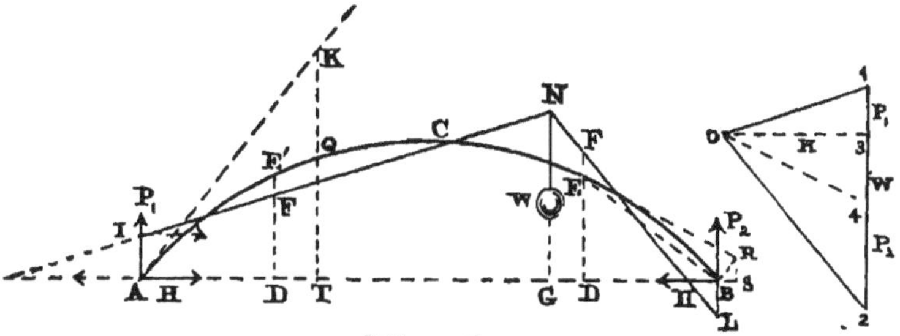

Fig. 3.

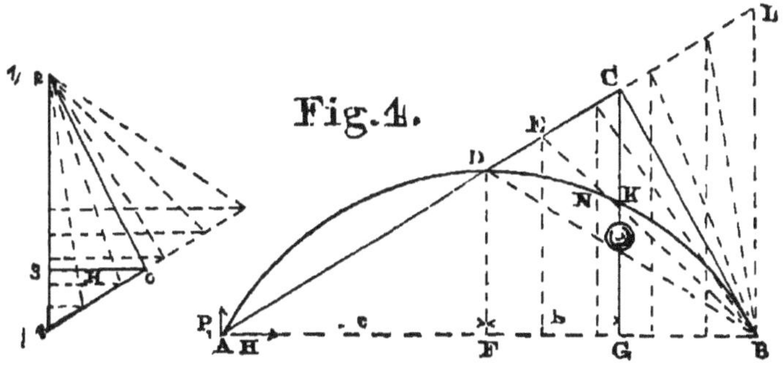

Fig. 4.

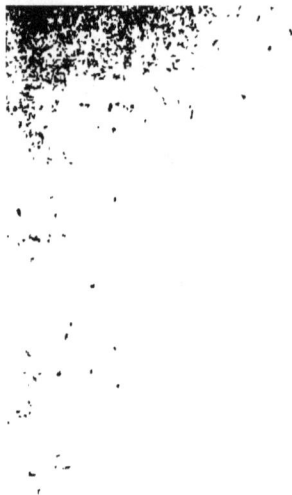

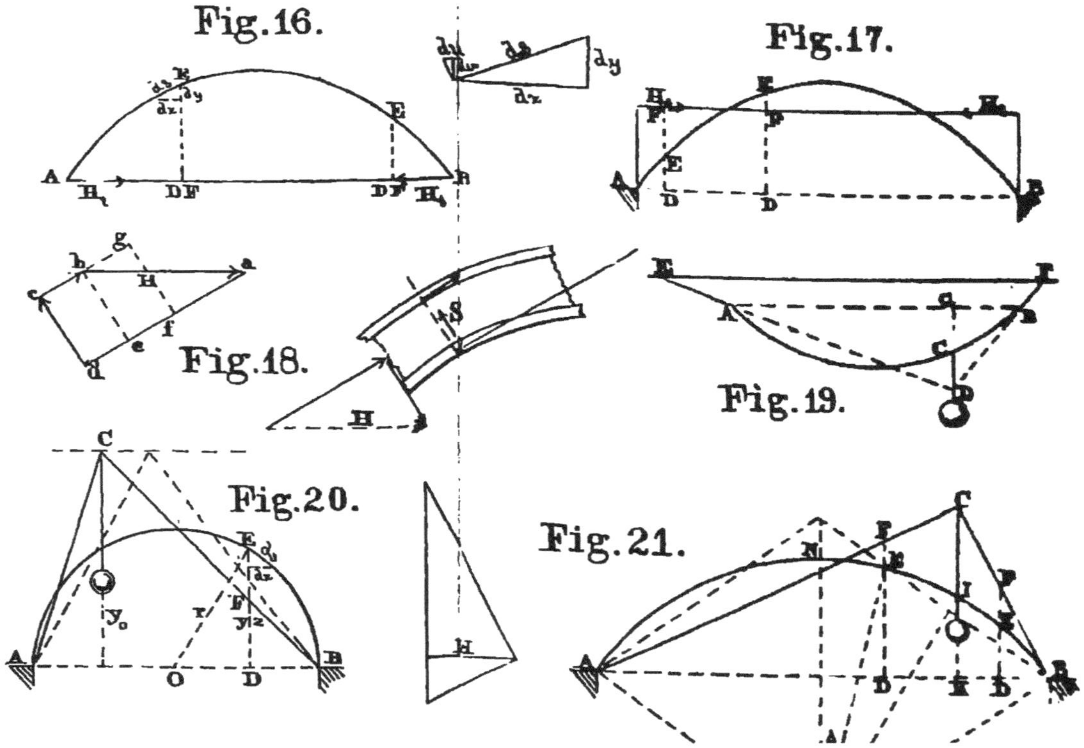

Fig. 16.

Fig. 17.

Fig. 18.

Fig. 19.

Fig. 20.

Fig. 21.

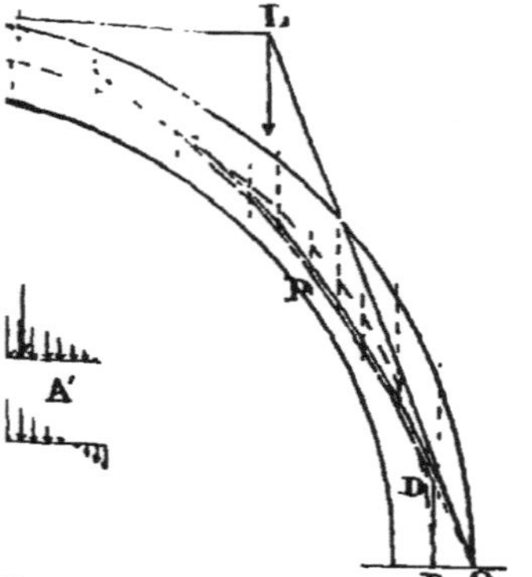

Fig. 26.

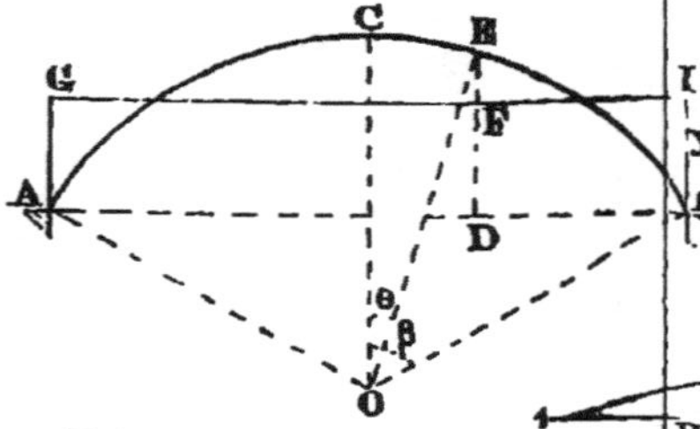

Fig. 27.

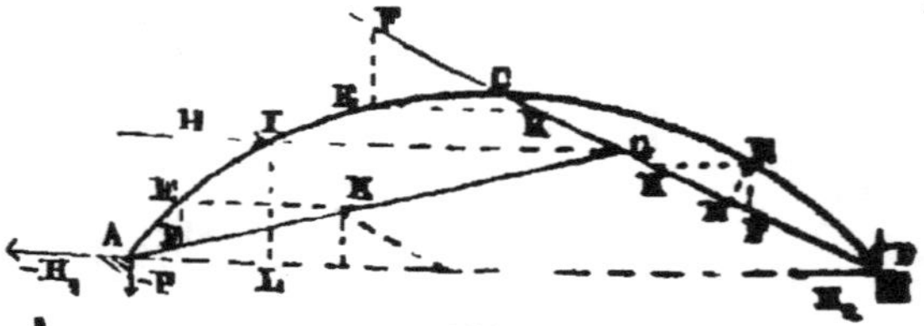

Fig. 32.

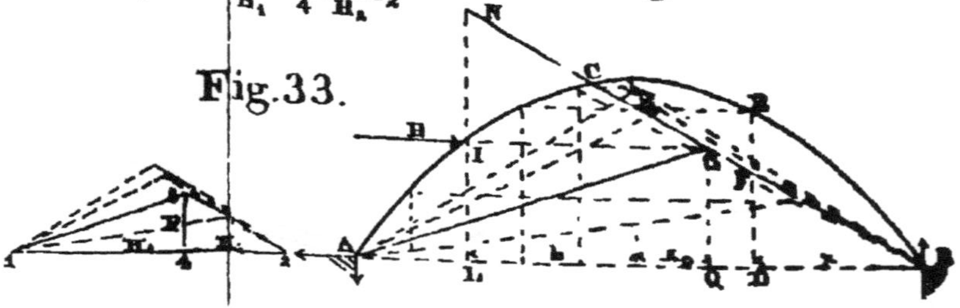

Fig. 33.

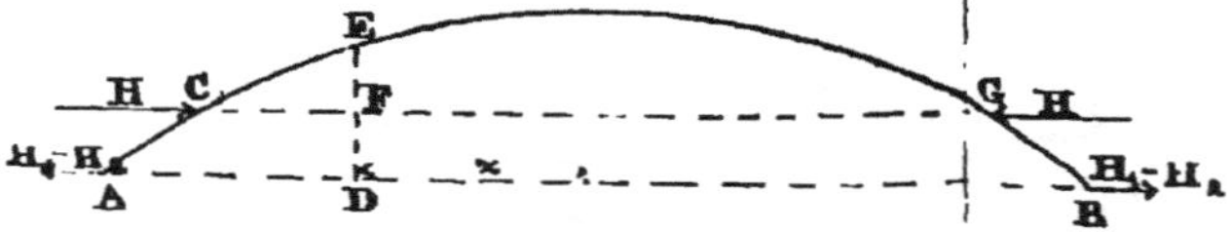

Fig. 34.

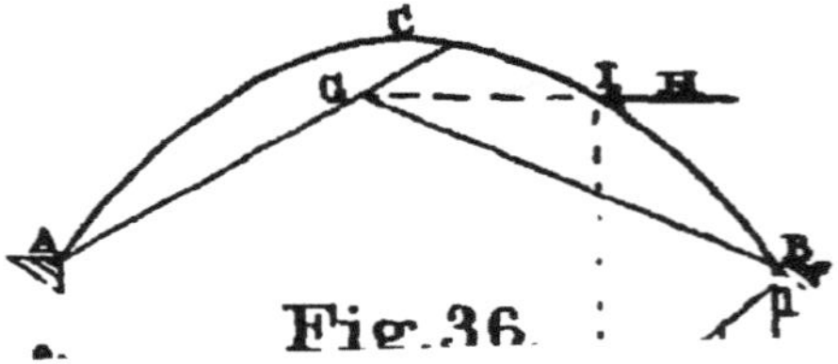

Fig. 36.

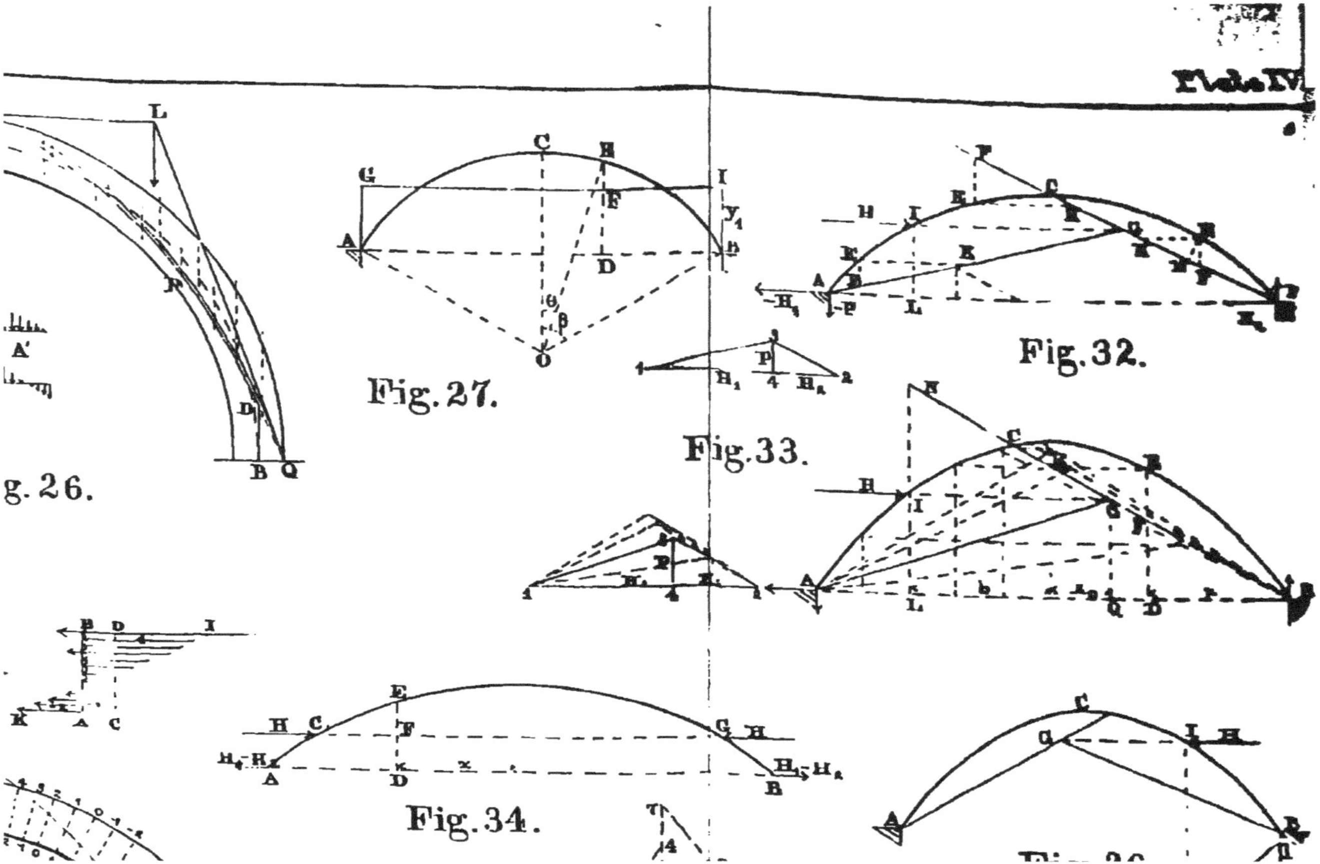

g. 26.

Fig. 27.

Fig. 32.

Fig. 33.

Fig. 34.

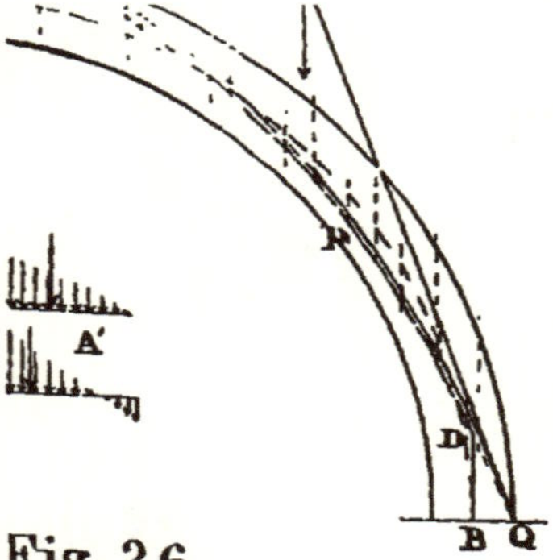

Fig. 26.

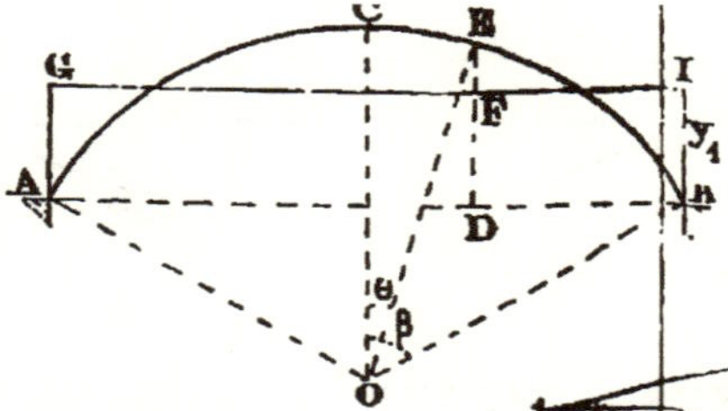

Fig. 27.

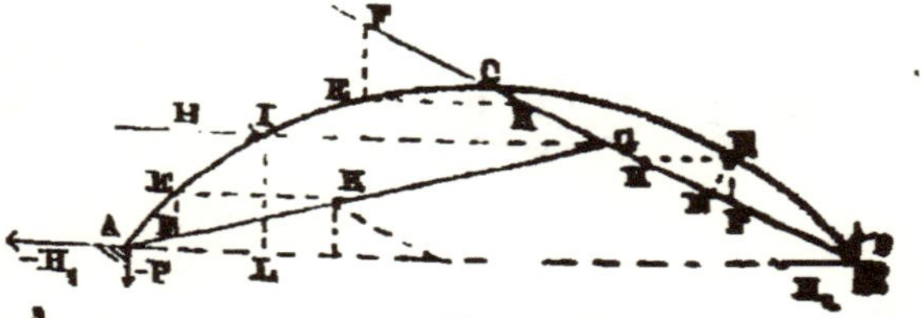

Fig. 32.

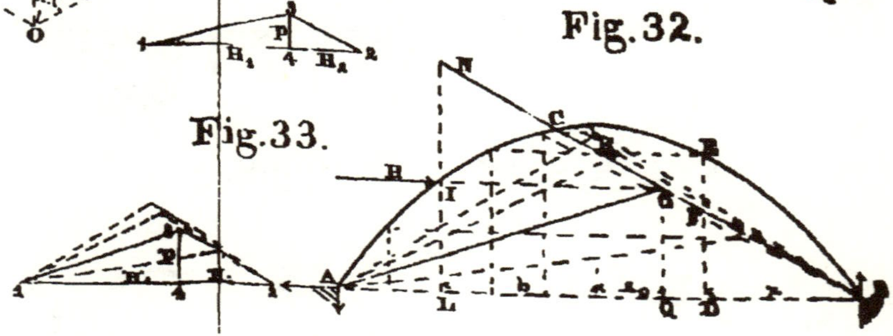

Fig. 33.

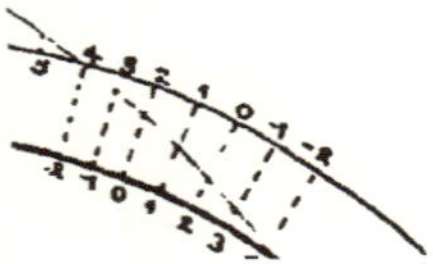

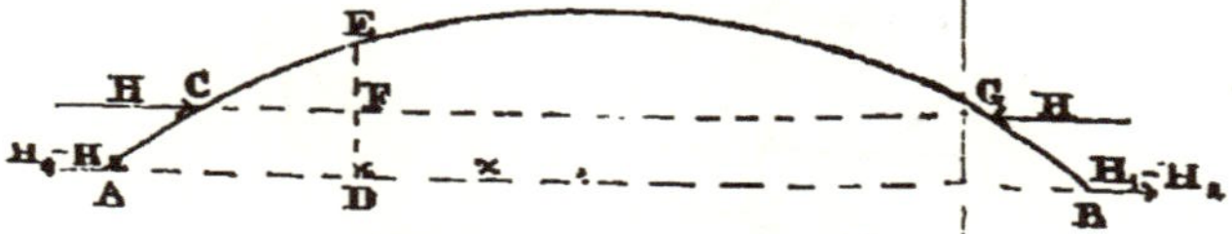

Fig. 34.

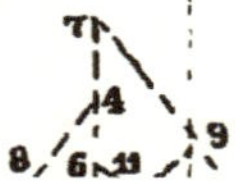

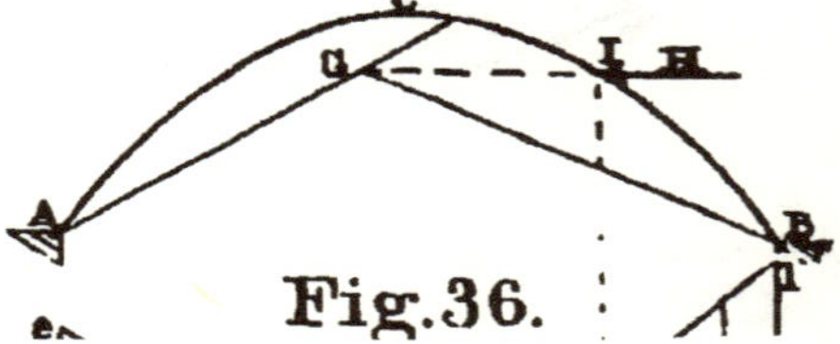

Fig. 36.

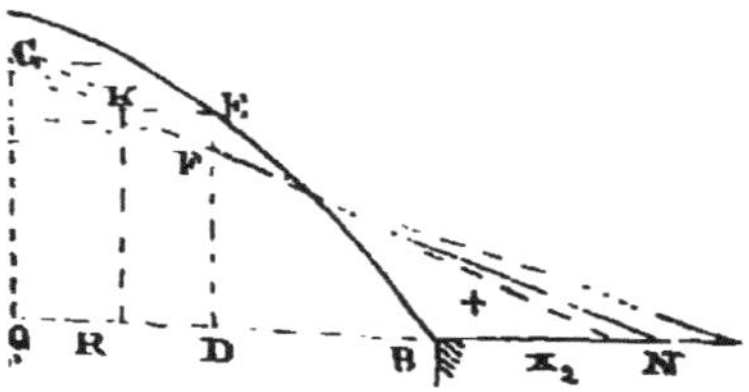

Fig. 38.

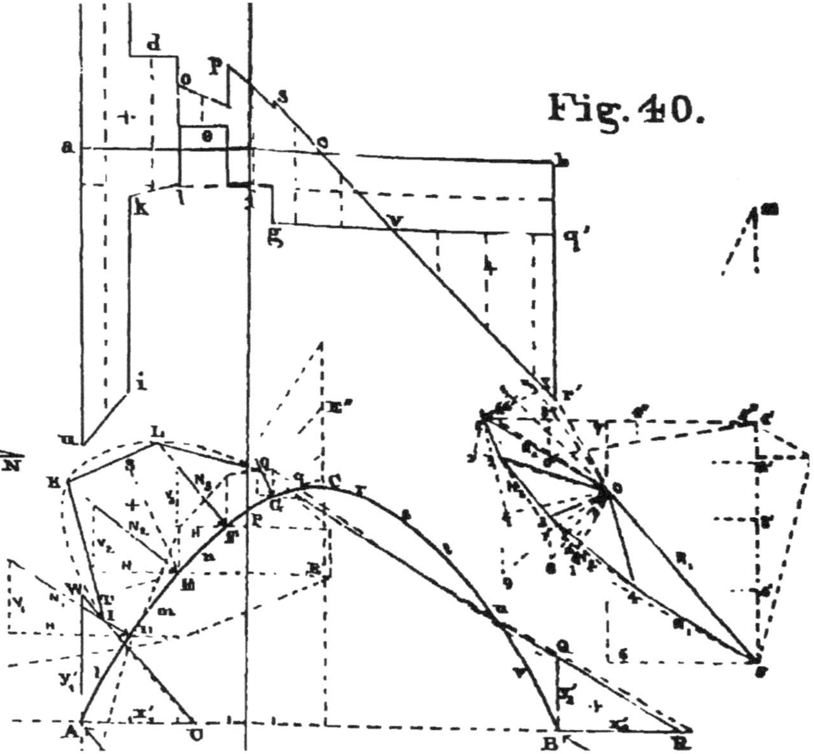

Fig. 40.

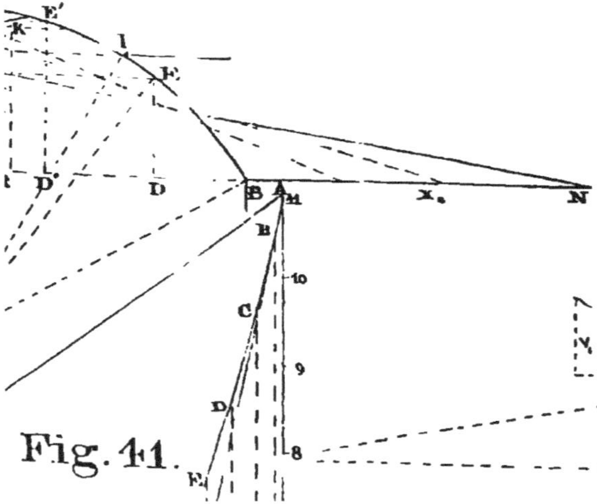

Fig. 41.

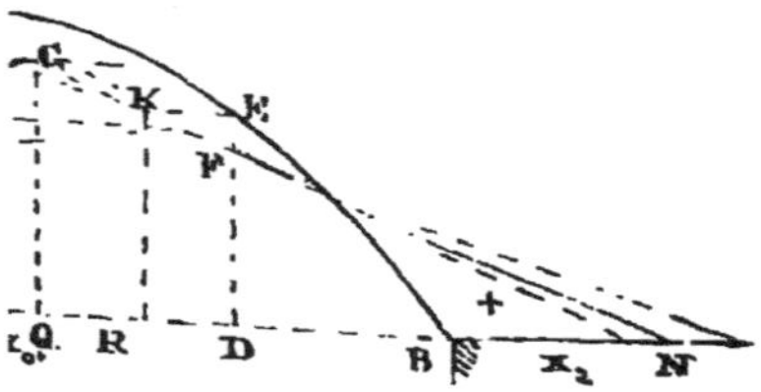

Fig. 38.

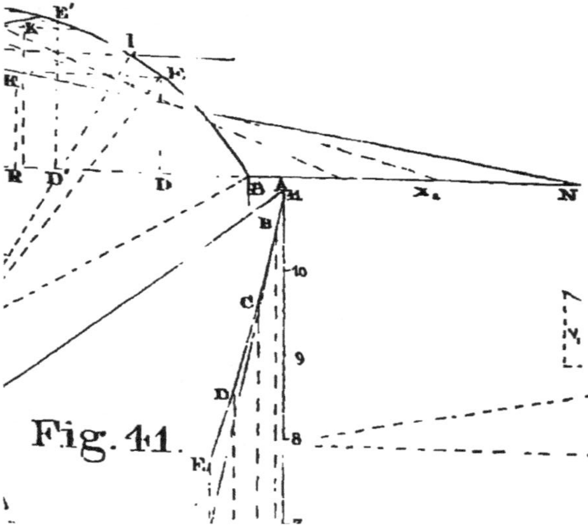

Fig. 41.

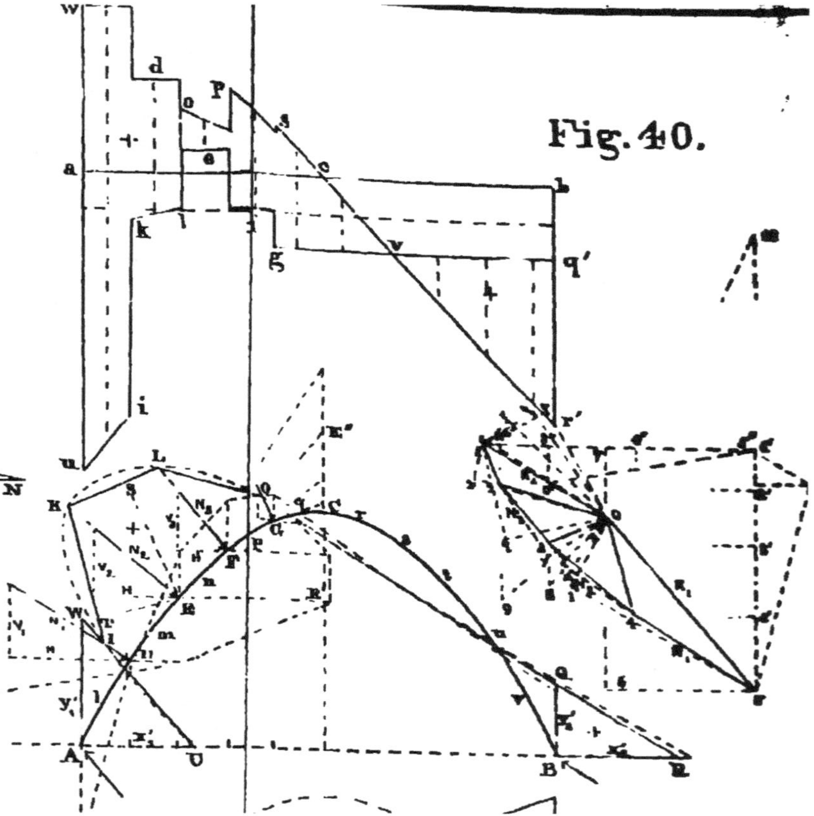

Fig. 40.

Fig. 44.

Plate VII

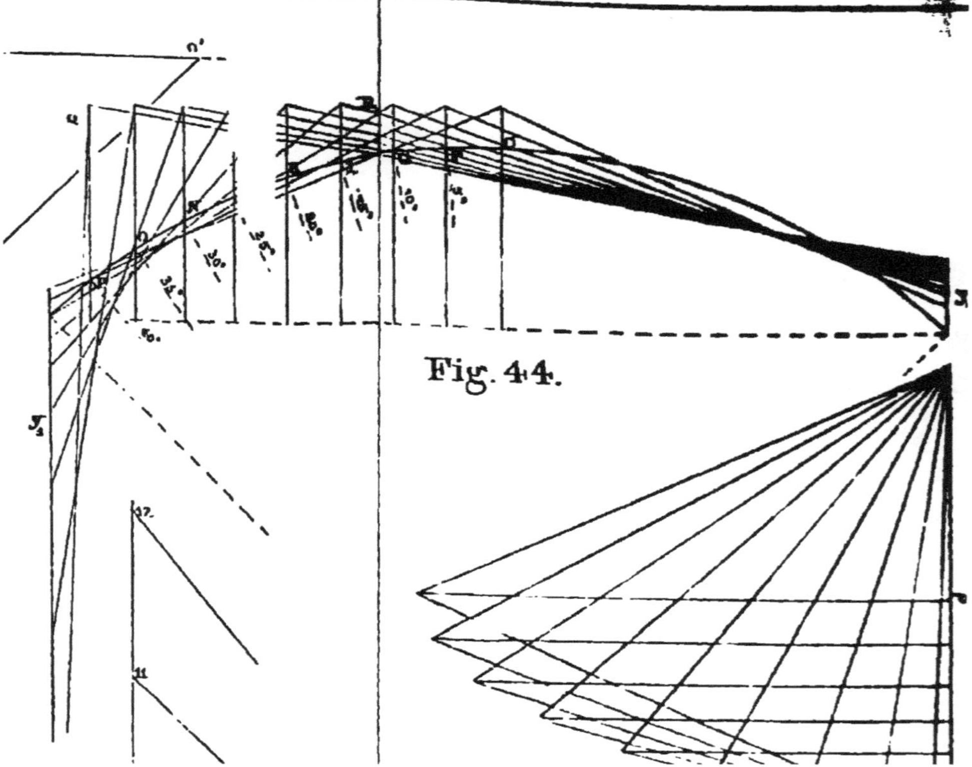

Fig. 44.

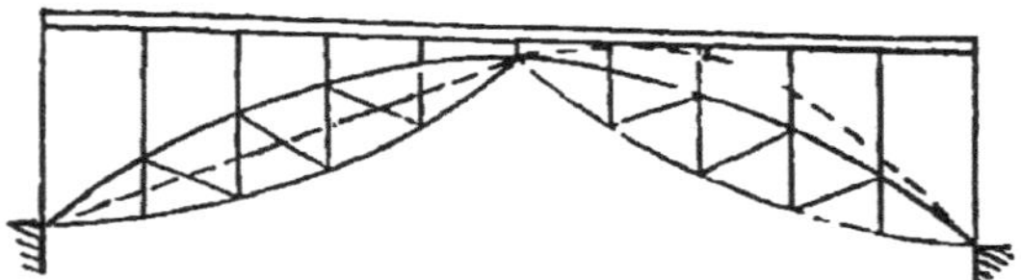

Fig. 5

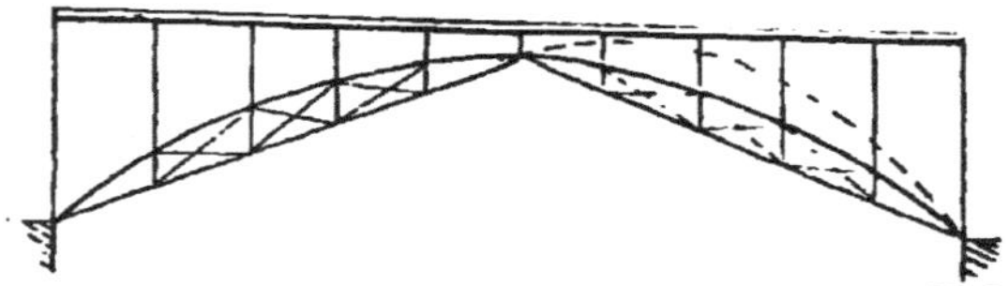

Fig. 6

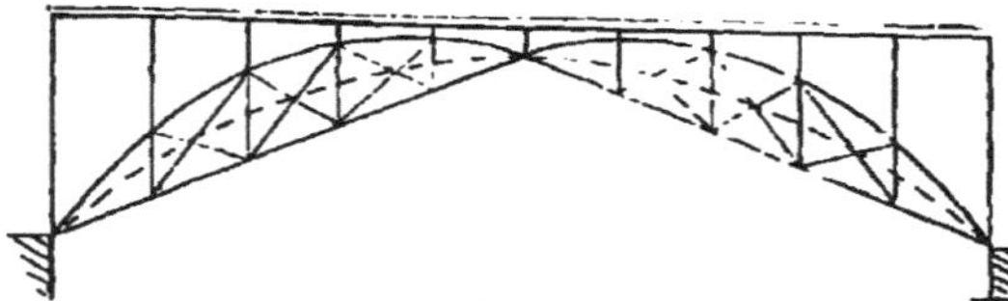

Fig 54.

Fig. 5

Plate VII

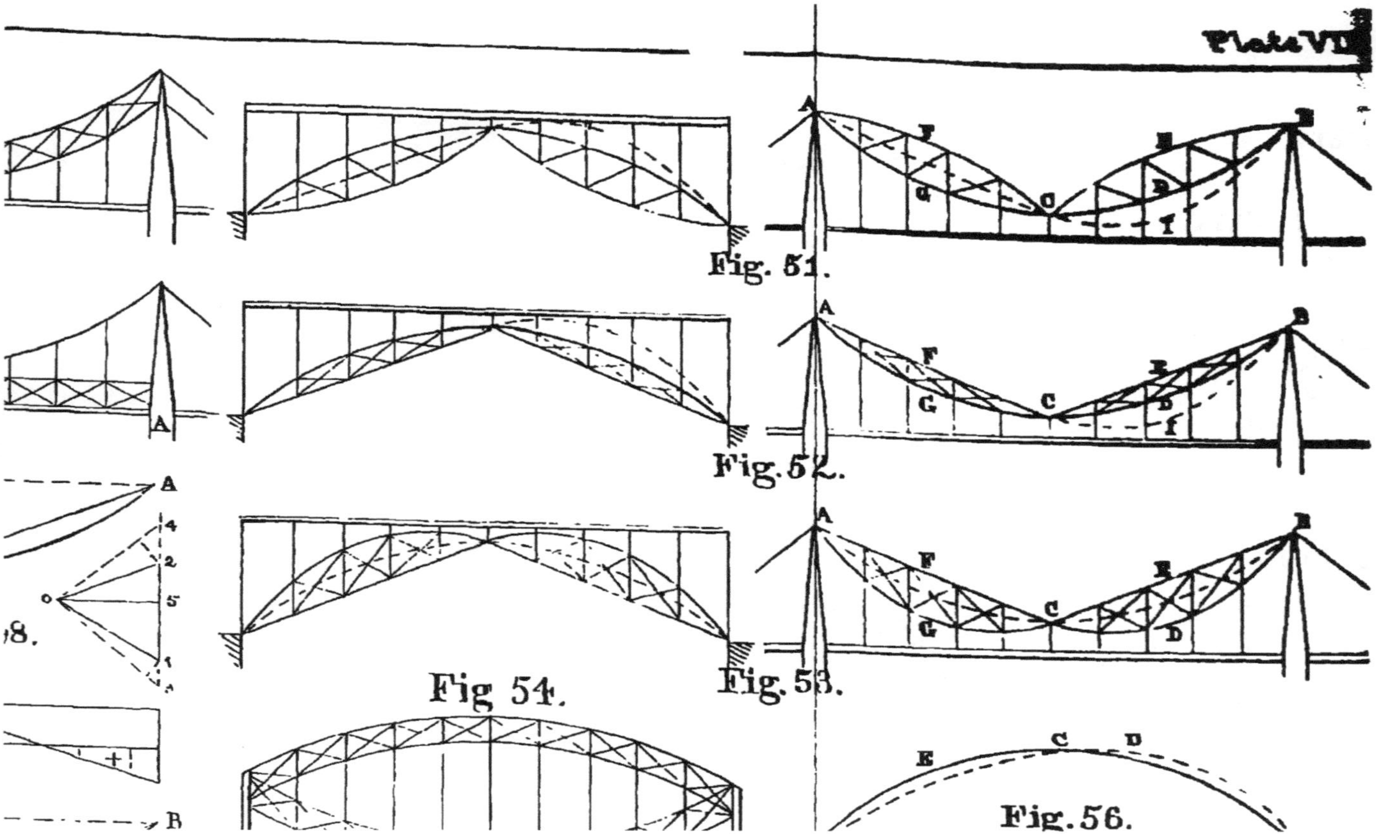

Fig. 51.

Fig. 52.

Fig. 53.

Fig 54.

Fig. 56.

8.

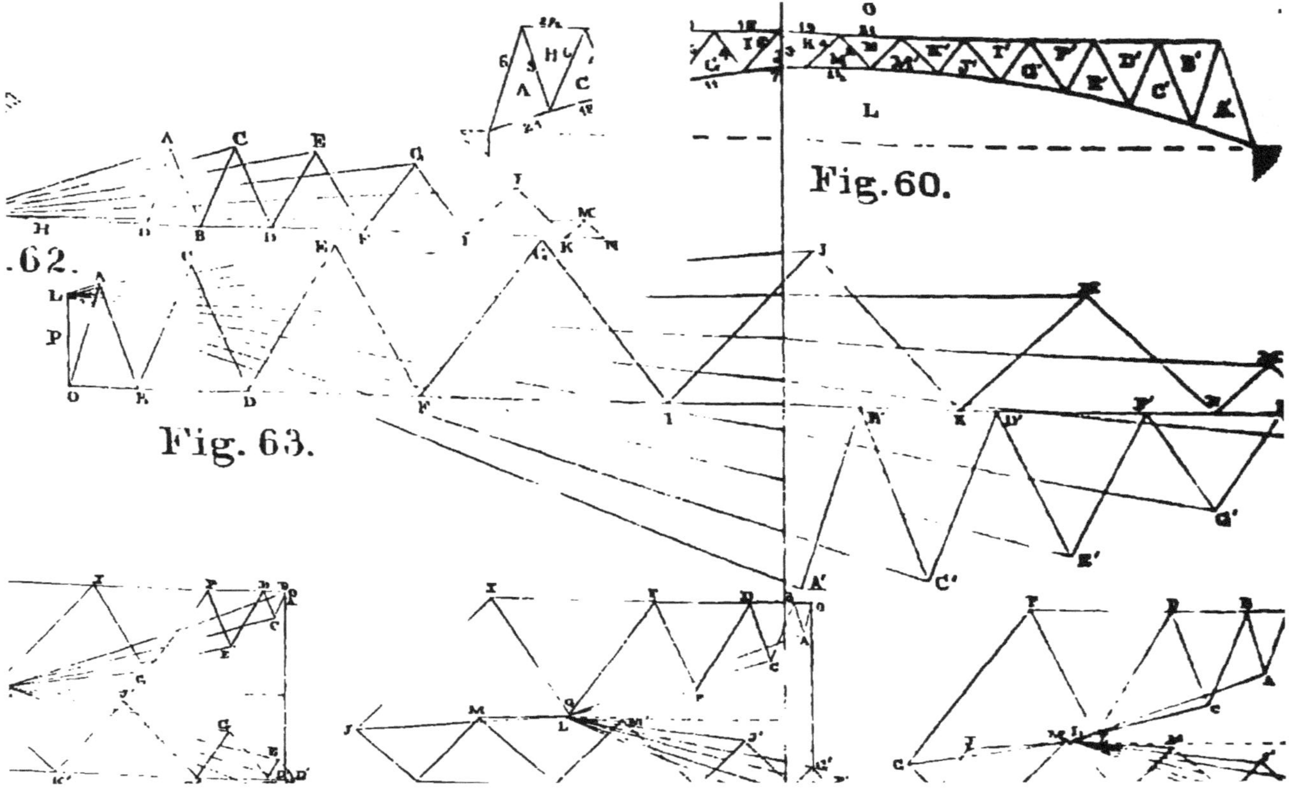

.62.

Fig. 63.

Fig. 60.

Plate VIII.

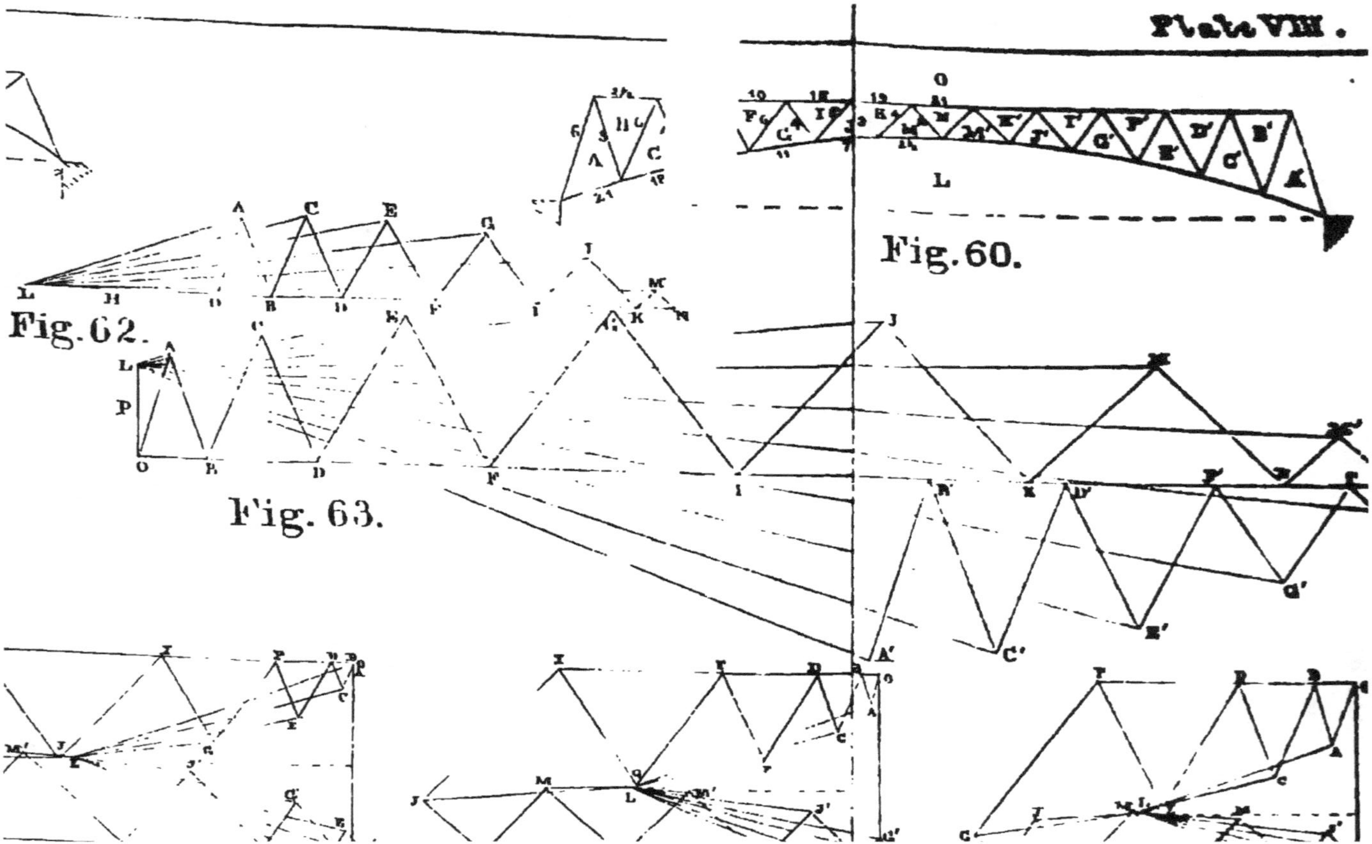

Fig. 60.

Fig. 62.

Fig. 63.

BRIDGES, ROOFS, Etc.

CANTILEVER—HIGHWAY—SUSPENSION.

PRACTICAL TREATISE ON THE CONSTRUCTION OF IRON HIGHWAY BRIDGES.

R. For the use of Town Committees, together with a short Essay upon the application of the principles of the Lever to a ready analysis of the strains upon the more customary forms of Beams and Trusses. With many fine wood engravings. By A. P. Boller, A.M., C.E. Fourth edition..8vo, cloth, $2 00

"It is written in plain, untechnical language, and presents rules for guidance, specifications, etc., that will be of inestimable service to town committees."

***THE THAMES RIVER BRIDGE.**

R. A Report to the General Manager of the New York, Providence and Boston Railroad upon the construction of the Thames River Bridge and Approaches at New London, Conn. By Alfred P. Boller, Chief Engineer. Illustrated with numerous folding plates and a handsome heliotype of the bridge.
Limited edition, 4to, paper, 5 00

A COURSE ON THE STRESSES IN BRIDGES AND ROOF TRUSSES, ARCHED RIBS, AND SUSPENSION BRIDGES.

Prepared for the Department of Civil Engineering at the Rensselaer Polytechnic Institute. By Prof. W. H. Burr. Eighth edition, revised. With Appendix on Cantilevers. Nearly the entire section of Swing Bridges has been completely rewritten and considerably extended. Plates, 8vo, cloth, 3 50

"No better practical work on Bridge Stresses has yet appeared."—*Mechanical World* (London).

"The book will be valuable not only to the student of Bridge Engineering, but to the Engineer who is already in practice."—*Journal Railway Appliances.*

A TEXT-BOOK ON ROOFS AND BRIDGES.

IMAN. Part I. Stresses in Simple Trusses. By Prof. Mansfield Merriman, Lehigh University. Third edition......8vo, cloth, 2 50

"I like the looks of your 'Roofs an[illegible] B[illegible] [illegible]ch, and expect to introduce it next term."—Prof L. M. H[illegible] *Penn.*

"The author gives the most modern practice in determining the stresses due to [illegible] loads, and reproduces th[illegible] eel movements. The whole treatment is concise and very clear and elegant."—*Railroad Gazette.*

A TEXT-BOOK ON ROOFS AND BRIDGES. Part II.

IMAN. ACOBY. Graphic Statics by Mansfield Merriman and Henry S. Jacoby. Second edition.............8vo, cloth, 2 50

"The plan of this book is simple and easily understood; and as the treatment [illegible], mathematics can scarcely be said to enter into [illegible] om our own correspondence, it is a work for which there is a decided demand outside of technical schools."
—*Engineering News.*

ROOFS AND BRIDGES. Part III. Bridge Design. By

RIMAN. Mansfield Merriman, Professor of Civil Engineering, Lehigh University. 8vo, cloth. *Shortly.*

MECHANICS OF THE GIRDER.

ORE. A Treatise on Bridges and Roofs, in which the necessary and sufficient weight of the structure is calculated. not assumed, and the number of Panels and height of Girder that render the Bridge weight least for a given Span, Live Load, and Wind Pressure are determined. By John D. Crehore, C.E. Illustrated by over 100 engravings, with tables, etc.....8vo, cloth, 5 00

"The Mechanics of the Girder for all the various shapes that it assumes before the Engineer, seems to have received here thorough and elegant treatment."—*Journal* [illegible]

BRIDGES, ROOFS, Etc.

CANTILEVER—HIGHWAY—SUSPENSION.

PRACTICAL TREATISE ON THE CONSTRUCTION OF IRON HIGHWAY BRIDGES.

R. For the use of Town Committees, together with a short Essay upon the application of the principles of the Lever to a ready analysis of the strains upon the more customary forms of Beams and Trusses. With many fine wood engravings. By A. P. Boller, A.M., C.E. Fourth edition..8vo, cloth, $2 00

"It is written in plain, untechnical language, and presents rules for guidance, specifications, etc., that will be of inestimable service to town committees."

***THE THAMES RIVER BRIDGE.**

IR. A Report to the General Manager of the New York, Providence and Boston Railroad upon the construction of the Thames River Bridge and Approaches at New London, Conn. By Alfred P. Boller, Chief Engineer. Illustrated with numerous folding plates and a handsome heliotype of the bridge.
Limited edition, 4to, paper, 5 00

A COURSE ON THE STRESSES IN BRIDGES AND ROOF TRUSSES, ARCHED RIBS, AND SUSPENSION BRIDGES.

Prepared for the Department of Civil Engineering at the Rensselaer Polytechnic Institute. By Prof. W. H. Burr. Eighth edition, revised. With Appendix on Cantilevers. Nearly the entire section of Swing Bridges has been completely rewritten and considerably extended. Plates, 8vo, cloth, 3 50

"No better practical work on Bridge Stresses has yet appeared."—*Mechanical World* (London).

"The book will be valuable not only to the student of Bridge Engineering, but to the Engineer who is already in practice."—*Journal Railway Appliances.*

A TEXT-BOOK ON ROOFS AND BRIDGES.

IMAN. Part I. Stresses in Simple Trusses. By Prof. Mansfield Merriman, Lehigh University. Third edition......8vo, cloth, 2 50

"I like the looks of your 'Roofs and Bridges' very much, and expect to introduce it next term."—Prof L. M. HAUPT, *University of Penn.*

"The author gives the most modern practice in determining the stresses due to moving loads, taking actual typical locomotive wheel loads, and reproduces the Phœnix Bridge Co's diagram for tabulating wheel movements. The whole treatment is concise and very clear and elegant."—*Railroad Gazette.*

A TEXT-BOOK ON ROOFS AND BRIDGES. Part II.

IMAN, Graphic Statics by Mansfield Merriman and Henry S. Jacoby.
ACOBY. Second edition.............................8vo, cloth, 2 50

"The plan of this book is simple and easily understood; and as the treatment of all problems is graphical, mathematics can scarcely be said to enter into its composition. Judging from our own [illegible] it is a work for which there is a decided demand outside of [illegible]."

—*Engineering News*

ROOFS AND BRIDGES. Part III. Bridge Design. By

RIMAN. Mansfield Merriman, Professor of Civil Engineering, Lehigh University. 8vo, cloth. *Shortly.*

MECHANICS OF THE GIRDER.

ORE. A Treatise on Bridges and Roofs, in which the necessary and sufficient weight of the structure is calculated, not assumed, and the number of Panels and height of Girder that render the Bridge weight least for a given Span, Live Load, and Wind Pressure are determined. By John D. Crehore, C.E. Illustrated by over 100 engravings, with tables, etc......8vo, cloth, 5 00

"The Mechanics of the Girder for all the various shapes that it assumes before t[illegible]ived here thorough and elegant treatment."—[illegible]

GREENE.

GRAPHICS FOR ENGINEERS, ARCHITECTS, AND BUILDERS.

A Manual for Designers, and a Text-Book for Scientific Schools.

TRUSSES AND ARCHES. Analyzed and Discussed by Graphical Methods by Chas. E. Greene, Prof. of Civil Engineering, University of Michigan. In THREE PARTS.

Part I. ROOF TRUSSES. Diagrams for Steady Load, Snow, and Wind. New revised edition (1890)..........8vo, cloth, $1 25

"This new edition of the first part of Prof. Greene's work on Graphical Statics contains some considerable additions, modifications, and rearrangements of material, tending to further improve the work, our favorable opinion of which is sufficiently indicated by the fact that the substance of the work is a reprint of a series of articles originally contributed to this journal."—*Engineering News.*

Part II. BRIDGE TRUSSES. Single, Continuous, and Draw Spans; Single and Multiple Systems; Straight and Inclined Chords. New revised edition, 1891...............8vo, cloth, 2 50

Part III. ARCHES IN WOOD, IRON, AND STONE. For Roofs, Bridges, and Wall Openings; Arched Ribs and Braced Arches; Stresses from Wind and Change of Temperature. Second edition........................8vo, cloth, 2 50

"So eminently simple as to be exactly fitted for working Architects and Builders."—Prof. Geo. L. Vose.

"We can recommend Prof. Greene's book as particularly adapted to students."—*Engineering News.*

"An excellent little manual which we can decidedly recommend." *Engineering (London).*

WADDELL.

THE DESIGNING OF ORDINARY IRON HIGHWAY BRIDGES.

A new Practical Work, with many Tables and Illustrations. Second edition. With an additional plate, etc. By J. A. L. Waddell, Member of the Society of Engineers, and Professor in the University of Tokio, Japan. Fifth edition....8vo, cloth, 4 00

"His book is probably the most valuable contribution to the literature of Iron Bridge Building which has yet appeared."—*American Engineer.*

WOOD.

A TREATISE ON THE THEORY OF THE CONSTRUCTION OF BRIDGES AND ROOFS.

Designed as a Text-book and for Practical Use. By Prof. De Volson Wood. Illustrated with numerous wood engravings. Seventh edition, revised and corrected.............8vo, cloth, 2 00

This work treats of all the well-known forms of Bridges, and several forms of Roofs. The analysis is generally of a simple kind, the essential parts of the work requiring only a knowledge of Algebra and Trigonometry. The examples are original and many of them novel in character. The latter part of the work develops the general mode of investigation by means of the general equations of statics.

FOSTER.

WOODEN TRESTLE BRIDGES.

According to the Present Practice on American Railroads, Treating of Pile Bents, Pile Drivers, Framed Bents, Floor System, Bracing Trestles of all Kinds, Iron Details, Connection with Embankment and Protection against Accidents, Field Engineering and Erection, Preservation and Standard Specifications, Bills of Material, Records and Maintenance, Working Drawings. By Wolcott C. Foster, C E 4to, cloth, 5 00

"The result is a book the like of which does not exist in any language, and which is often called for by practising engineers and also in technical schools."—*Railroad Gazette.*

JOHNSON.
BRYAN.
TURNEAURE

THEORY AND PRACTICE IN THE DESIGNING OF MODERN FRAMED STRUCTURES.

4to, cloth, 10 00

www.ingramcontent.com/pod-product-compliance
Lightning Source LLC
Chambersburg PA
CBHW021705210326
41599CB00013B/1534

* 9 7 8 3 3 3 7 3 4 9 4 6 2 *